Sun Dancing

ALSO BY GEOFFREY MOORHOUSE

The Other England

The Press

Against All Reason

Calcutta

The Missionaries

The Fearful Void

The Diplomats

The Boat and the Town

The Best-Loved Game

India Britannica

Lord's

To the Frontier

Imperial City: The Rise and Rise of New York

At the George (essays)

Apples in the Snow

Hell's Foundations: A Town, Its Myths and Gallipoli

Om: an Indian pilgrimage

A People's Game

Sun Dancing

A VISION OF
MEDIEVAL IRELAND

Geoffrey Moorhouse

A Harvest Book
Harcourt Brace & Company
SAN DIEGO   NEW YORK   LONDON

Library of Congress Cataloging-in-Publication Data
Moorhouse, Geoffrey, 1931–
Sun dancing: a vision of medieval Ireland/Geoffrey Moorhouse.—
1st ed.
p.   cm.
Includes bibliographical references.
ISBN 0-15-100277-0
ISBN 0-15-600602-2 (pbk.)
1. Monastic and religious life—Ireland—History—Middle Ages, 600–1500—Fiction.
2. Monastic and religious life—Ireland—History—Middle Ages, 600–1500.
3. Skellig Michael (Monastery: Ireland)—History—Fiction. 4. Skellig Islands (Ireland)—
History—Fiction. 5. Skellig Islands (Ireland)—Antiquities. 6. Civilization, Medieval—
Fiction. 7. Civilization, Celtic—Fiction. 8. Ireland—Civilization.   I. Title.
PR6063.0655S86   1997
823'.914—dc21   97-8748

Text set in Granjon
Designed by Lori McThomas Buley
Printed in the United States of America
First Harvest edition 1999
C E F D B

To my Irish friends,
on both sides of the water.
May peace be upon us all

CONTENTS

Author's Note   IX

PART ONE
The Tradition

ONE   AD 588   3

TWO   AD 670   19

THREE   AD 780   35

FOUR   AD 824   55

FIVE   AD 950   71

SIX   AD 1044   85

SEVEN   AD 1222   105

PART TWO
The Evidence

Fionán   119

Peregrinatio   122

Pagans and Christians   126

St. Patrick   130

Brendan the Navigator   136

Irish Monasticism   143

Book Illustration   146

The Scoti   154

Antony and Cassian   156

The Desert Fathers   159

St Antony in Art   167

The Irish Kings and Tara   173

Three Martyrdoms   177

The Curach   182

Fasting 185

Monastic Dress 190

The Gallican Creed 192

Skellig Michael 195

Nature Worship 198

The World View 200

The Buildings 202

Soul-friends 205

The Monastic Horarium 208

Cú Chulainn 212

Two Tonsures 214

The Synod of Whitby 216

Columcille and Iona 219

Mingled Traditions 222

Monastic Feuds 223

Penance 225

Sin 230

Poverty and Sickness 232

The Vikings 236

Celtic Metalwork 240

Skellig Birdlife 243

The Dancing Sun 245

The Irish Annals 247

Brian Boru 249

Olaf Tryggvason 251

The Culdees 254

Giraldus Cambrensis 257

The Hermitage 259

Origen 261

The Bestiaries 263

Hubris 266

The Withdrawal 268

St Benedict's Rule 273

Ballinskelligs and the
  Arroasians 275

Unanswered Questions 278

A Bibliography 281

AUTHOR'S NOTE

This book has been a long time coming together: thirty years at least, and probably more. There was a holiday on the west coast of Ireland, when my children were very young and in no mood for anything but messing about on a sandy beach, or investigating the rock pools left by an ebbing tide. Climbing a headland on a day of newly rinsed clarity, I noticed for the first time two jagged shapes far out to sea, where they had earlier been hidden by rain-cloud or fog. There was something about their outlines that made them seem untouchable, made me think of them immediately as rocks and not as islands. A little later, I asked an old man about them and his eyes widened, his brow lifted, as he told me that I had seen the Skelligs, and would be forever restless until I had set foot on one of them. 'For that is holy ground over there,' he said, explaining that monks had dwelt on Skellig Michael for hundreds of years, while its companion, Little Skellig, had always been a hallowed sanctuary of birds.

Restless I remained for many more years, because Skellig Michael can only be gained across a lenient sea. The day may seem beautiful to people ashore, but if any kind of a swell is running no boatman will waste his time going over from the mainland, when he knows that a landing is impossible in the heave and fall of an Atlantic surge. Three more times I was to stand frustrated on that western coast, with long intervals in between, before at last I stepped onto the greater of the two rocks, beneath an almost vertical cliff. It was 1986 and I was lecturing on a cruise vessel that was circumnavigating the British and Irish Isles. The ship was owned by an American company whose proud boast was that there was nowhere its passengers could not land; which was effectively true, when it put them afloat in powerful inflatables, driven by gung-ho young men whose fathers had doubtless done similar service in the Marines at Iwo Jima and Guadalcanal.

As I climbed the path winding up to the ancient constructions near the top of that cliff, I sensed that I was on the threshold of something utterly unique, though I was by no means a stranger to monasteries, which I had visited throughout Europe and even further afield at one time and another. But nothing in my experience had prepared me for this huddle of domes, crouching halfway to heaven in this all but in-accessible place, with an intimidating immensity of space all around, where it was easy to feel that you had reached a limit of this world. A holy place, to be sure, which would have still been so, even if it had never known the consecrated life of prayer. I have returned to it many times since, and I have never lost the spellbound sense of awe I felt on that first marvellously quiet, sun-blessed day in spring. It was then that I knew I must write about the skeilic, try to fix its essence in the collective memory; but I never have, until now.

And this is a cautionary Note, for the benefit of those readers who like to know in advance what they are letting themselves in for. I would not dare to suggest that the structure of this book breaks com-pletely new ground, but it is unusual; even, perhaps, experimental: an attempt to reconstruct history from very limited direct historical evi-dence. Part One can be read as a fiction about a particularly intense form of the monastic life in medieval Ireland; or as a meditation on Celtic spirituality, which is as valid today as it was in the early Middle

Ages. Though the story told in these first pages draws heavily on the imagination, its basis lies in the historical reality of events which began to happen some fourteen hundred years ago, on an inhospitable rock which rises sensationally from the Atlantic.

Part Two can be seen as a continuation of the narrative, which amplifies the story in a different voice; or it can be used for reference—it is arranged so that the reader can easily turn to it at any time—by those impatient for further light to be cast on incidents as they first occur. Its range is much wider than the particular theme of Part One, and it enlarges in some detail on topics that are only just detectable beneath the surface of the earlier text. It treats of the roots and essences of Irishness, including attitudes to penance and fasting which now seem savage and which even in the Middle Ages were thought to be exceptionally rigorous; of connections between medieval Ireland and places as remote from it as India, Central Asia, Egypt and Byzantium; of Irish peregrinations that left a lasting mark across the length and breadth of continental Europe; of the early Desert Fathers and the effect that one of them had on the subsequent course of Western art; of the Celtic genius for amalgamating pagan naturism with Christian theology; of the progression from a biblical understanding of creation to a scientific apprehension of the universe; of medieval sickness and poverty and violence which have too often been coated in romantic humbug; of the effect on an isolated culture of Viking invasion, ecclesiastical Romanisation and Anglo-Norman settlement. Niall of the Nine Hostages, Cú Chulainn, Brian Boru, Patrick, Brendan, Columba and Brigid of Kildare cross these pages; so also do Paul of Thebes, Athanasius, Cassian, Antony the Hermit, Origen, any number of popes, Olaf Tryggvason, Hieronymus Bosch and Gustave Flaubert.

Sun Dancing is about a harsh and obsessed way of life on the very edge of the known world, which (and we have Sir Kenneth Clark's word for this) was partly responsible for the survival of Western civilisation during the Dark Ages of European history. It is an attempt to explain and illuminate that distant time, a chronicle full of paradoxes, a measurement of persistent belief in the supernatural, which can sometimes enlarge the human spirit, but can also turn it to madness. The book, however, ends with two explicit question marks, and many others are left hanging in the air. Some of the questions cannot be answered. Not yet, anyway.

PART ONE

The Tradition

As far as they can be verified, the events described in the following narrative took place between the sixth and the thirteenth centuries, off the west coast of Ireland

O N E

AD 588

The men at the oars moved as though they had all the time in the world at their disposal. They reached forward easily to dip the long thin spars into the tranquil sea, and leaned back almost lazily to pull their craft along. With three oarsmen on each side, this seemed to skim across the surface with as little effort as they had needed to come down the long sweeping bends of the river the day before. Six men more lay resting against the bags full of tools and other tackle, the water-skins and the packages of food, awaiting the moment when they would be told to change places by the gaunt figure in the stern. Where Fionán sat with arms outstretched along the wooden frame of the boat, a tightly wrapped bundle of the most precious things beside him, and enjoyed the sensation of being borne along to a great and mysterious destiny.

His bony face was tilted toward the sky, as if to invite yet more grace, and he closed his eyes against the glare, which warmed his features and the pale dome of his forehead. He was elated that the enterprise he had longed for years to begin was now under way, but was content that he could not even guess how it would finish and where it might end. Part of him could have wished for a breeze, so that they might have raised the little sail to ease their passage more, but he did not fret at the lack of it. The time spent in seeking their place of solitude across the sea was not theirs to ordain. It, too, was in the Maker's gift, with its own divine purposes, and He would settle the duration of this journey as well as its outcome, wherever that might be.

Their course had already been providentially determined before the moon rose on the day they set out. They had reached the estuary and, where the river met the sea, Fionán told the oarsmen to rest and wait, to find out which way their boat would drift. That had been Brendan's counsel, whispered urgently as he lay dying, his eyes burning as brightly as ever while he awaited his own last journey, and gave his final instructions to the man he had carefully fostered from youth.

'When you reach the sea,' he told Fionán, 'the moment the waves begin, there you must surrender to what will be and wait for the sign. Wherever the boat begins to lead, that way you must go from then on. It will be meant.'

They had waited for only a moment, expecting to be borne on towards the setting sun, when suddenly their craft, riding lightly on the very surface of the water like a feather in a pond, was caught by a puff of wind from the west, and its prow slewed sharply to the side. 'There lies our way,' said Fionán, pointing ahead. 'Our desert is down there.'

They therefore rowed on with the coast at their left hand until darkness fell, when they stopped to say their prayers, to take a little food and water, before hooding their heads and composing themselves to sleep. There was nothing to disturb them through the night, for the sea was very calm, so that they were soothed by its gentle motions, every ripple and eddy of which they could feel and hear through the hides, almost as clearly as if the water was touching their own skins. When dawn came, the land was still in sight, but more distantly now, some hills emerging indistinctly as the sun rose behind, above a haze. They had drifted away from the shore in the darkness, but not one of them felt even a twinge of anxiety, for this too was part of the carefully prepared plan. Once more six oars began to swing in unison and the boat raced along over the still-placid sea, causing some gulls to stride out of its path in alarm, wings flapping frantically until they had beaten a way into the sky. The good Lord, thought Fionán, was surely with them here, and for that he offered heartfelt thanks; but he also blessed his own instinct that had kept faith with the old tradition, so that they had deliberately set forth to find their destiny on the feast of Lughnasadh, when all was ripe and bountiful and benign upon the land, and when the deep waters scarcely breathed even gently so. This was almost the only decision that had not been forming in Fionán's mind for many years. At first he had thought the most propitious time might be on the occasion of Patrick's feast, but the weather had been so stormy then, for year after year, that he had abandoned it. It was not that he feared to put himself at risk, nor even because he would be re-sponsible for his brothers' lives, too; for these things were truly of lit-tle account, and not one of them did not long to see the divine face as soon as he could. But there was a purpose to perform first; the reason, Fionán believed, that he had been sent, and he could not hazard this. He was to be a sign.

He had discerned this vocation as a young man, shortly after joining the monastery. He had inherited his father's thirst for knowledge, as well as his joy in all things that were beautiful: in their colours, the textures, in the way they looked and felt and otherwise were. His fa-ther would speak as gently as any woman when he drew Fionán's at-tention to the brilliant sheen along the outstretched neck of a flying duck, or the constantly changing shape of a passing cloud, or the seething of the wind through the branches of a midsummer tree. It was a time of great peace in Munster, after long years of warfare be-tween the clans, and the craftsman relished the tranquillity that al-lowed him to work in wood and stone without distraction, and to reflect on the precious integrity of all created things. 'God is so good to us,' he would say, 'to surround us with so much loveliness.' He be-queathed his great piety, as well as his sensibility, to his children, and tenderly accompanied rather than coaxed his youngest son in the direction of the monastery, which the boy entered in his eighteenth year. It was there that Fionán discovered both Brendan and Antony, and afterwards knew exactly what he must do with the rest of his life.

Brendan was passing from middle to old age when Fionán became a monk, though he would live for many years more, quite long enough to have a lasting influence on the young man's spiritual and ascetic formation. Brendan's great wanderings across the seas were over, and he now regarded them as having been less a search for his own place of resurrection, which he had formerly believed, as a preparation to be a father in God not so very far from where his life began. An abbot who had satisfied his wanderlust, and who was wise enough to recognise each of the motives that had sent him restlessly away, might be better employed founding a school of the Lord's service nearer home. So he had discovered when lying prostrate before the altar one night.

Brendan became aware of Fionán's particular gifts shortly after his arrival in the new community, so that instead of tilling the soil or undertaking other forms of manual labour, as most of his brethren did when not at prayer or studying the written Word, the young man spent long hours in the scriptorium copying texts with three other monks, entrusted with the task of making the monastery's own library from originals that Brendan had cajoled from Clonmacnois. Fionán soon realised that there was nothing in the world that he would rather be doing than these daily acts of creation as a servant of the Lord; nothing that surpassed even the initial drudgery of producing the materials with which he would slowly and painstakingly fashion a book. His father had taught him well many of the skills that he now used every day. Others obtained calfskins and prepared the vellum for him, but it was Fionán who made his own goose quills and reeds to pen the texts of these manuscripts, his own brushes for their illumination with hairs from his own head, and from the hides of cattle and pigs. It was Fionán who painstakingly pulverised and mixed the minerals and the plants to produce the dyes that would light up a chapter, a paragraph, a page, a line: the red ochre from the earth, the green of verdigris, the yellow of orpiment, the golden-brown of lichen, and essences of blue from the honey-smelling woad (he ached to come by an even more lustrous blue, which was said to be made by grinding a precious stone, but that was found only in a land so far beyond Jerusalem that it was too remote to contemplate, and the stone was therefore much rarer than gold). It was Fionán who cooked fish for hours until his eyes stung with smoke, in order to extract the oil that would bind his colours to a page. Though he sometimes retched in the stink of other and even more noisome fats, it was he who would patiently burn them for the soot that made his inks, and to his surprise find that after a while he was as deeply absorbed in the process as in any other that was necessary for the creation of his precious volumes.

But nothing at all compared to the delight of actually copying the biblical and other words, which he did with small humphs of satisfaction as he penned an especially satisfying curve, or added a distinctive pothook to a character and later embellished it with a crimson stroke of the brush, all his own work as his confidence grew, and not a mute imitation of the Clonmacnois scribes. In two carefully planned columns each page was arranged, but now and then Fionán's calculations went astray, which meant that several words might have to be slipped in as unobtrusively as possible between existing lines; the head under the wing, or the turn in the path, as he and his companions knew it, and he was especially deft in that. In time he would draw a dainty little fish occasionally above the lettering of Christ's name, as a gratuitous embellishment to confirm his own anonymous artistry to any who might read the book in years to come. But stopped doing so one day, and never repeated this vanity, when it was pointed out to him that he was suffering from one of the deadliest sins.

He had come to the monastery with a modicum of Latin, picked up in his family's prayers and under the further encouragement of his father. Now he quickly learned more, both in concert with his brethren, as they recited the psalms and made other devotions together each day; still more in his liberating and transforming access to books, including a textbook which Brendan acquired to further the development of this promising boy. The monks of Clonmacnois had been generous in opening their own library to the younger community, and it was understood that Fionán and his companions might have anything they sought for as long as it took to copy it. They would, of course, always be guided by their abbot, and a start was therefore made with the Gospels, which took them nearly two full years to complete. After that they began the long reproduction of the older texts, of psalters and missals. Only then, ten years after they had embarked on this task, did Brendan direct them towards the manuscripts that were to alter the course of Fionán's life. It was in these that he discovered Antony.

There were two patristic volumes on the shelves of Clonmacnois that revealed to Fionán things about the beginnings of the faith he had never come across before. He knew all about the great conversion of the land that had been wrought generations ago by the redeemed slave in whose honour his own eldest blood brother had been baptised Padraig, and he was aware that once before the return of the saint, so it was said, a man sent from God had crossed the sea from Gaul with the message that most people had then been unable to believe or were unwilling to receive. It was even conjectured that a Scotus, one of their own people, had been present at the Crucifixion, and had made his way home across the seas to bear witness to this. But between the time of Paul and the other Apostles, and the era of the great conversion, the history of the faith was very nearly a blank as far as Fionán was concerned. Avidly he now fell upon the two texts in which he anticipated, the moment he set eyes on them, a greater enlightenment. One was a spiritual commentary on the early Fathers of the Church, the other a Life of Antony the Hermit, the two books written by different but equally enthusiastic authors.

Both portrayed a world in which men of faith went out into the deserts of Egypt, sometimes to a life of absolute solitude, sometimes as a small band of brothers, in order to be closer to God. That was the point. They were not fleeing from something they disliked or mistrusted or were polluted or persecuted by, though some of them had been persecuted, like all believers at that time. They were not taking refuge in the desert, as someone without faith might see it, out of fear of the alternative. They desired more than anything else to see God, to behold Christ's face, to know Him as He was, now and forever, awaiting them in heaven above. They were impatient for death, though this last and greatest blessing could only be granted them in God's good time and not at a moment of their own choosing, which would have been blasphemy. There was nothing else that mattered to them other than the desire for heavenly union, and they wished to prepare for it, to purify themselves and thus be worthy, to make it more certainly theirs, without the slightest distraction. That was all they were fleeing from: the distractions of the villages and towns that would hold them back from more than the possibility of God's grace. They did want much more than the possibility.

None was more ardent than Antony, and by the end of his life none had become more of a perfect example to others who would take this road. But it was not an easy path that he had chosen: so much was clear from the start, for the Devil appeared to tempt him in the guise of a woman and, when that failed to deflect him from his chosen purpose, had returned and 'approaching one night with a multitude of demons he whipped him with such force that he lay on the earth, speechless from the tortures'. Other demons tormented him with the appearance of lions, bears, leopards, bulls, serpents, asps, scorpions, wolves, 'and each of these moved in accordance with its form', striking him anew at will, and causing him to suffer yet more agony. He was not abandoned by the Lord, however, in this travail. 'For when he looked up he saw the roof being opened, as it seemed, and a certain beam of light descending towards him. Suddenly the demons vanished from view, the pain of his body ceased instantly, and the building was once more intact ... Antony entreated the vision that appeared, saying "Where were you? Why didn't you appear in the beginning, so that you could stop my distresses?" And a voice came to him: "I was here, Antony, but I waited to watch your struggle. And now, since you persevered and were not defeated, I will be your helper forever, and I will make you famous everywhere."'

Strengthened still more, Antony found a building beside the river, so long deserted that it was full of reptiles; nevertheless he decided to settle there, at which 'the creeping things departed, as if someone were in pursuit . . .' For twenty years Antony lived there in solitude, 'not venturing out and only occasionally being seen by anyone', his daily bread being brought to him twice a year by good people who marvelled at his resolution and his stamina. They marvelled even more when finally he emerged from his anchor-hold and it was seen

that his body had maintained its former condition, neither fat from lack of exercise nor emaciated from fasting and combat with demons . . . The state of his soul was one of purity, for it was not constricted by grief, nor relaxed by pleasure, nor affected by either laughter or dejection . . . He maintained utter equilibrium, like one guided by reason and steadfast in that which accords with nature . . . Thus he consoled many who mourned, and others hostile to each other he reconciled in friendship, urging everyone to prefer nothing in the world above the love of Christ. And when he spoke and urged them to keep in mind the future good and the affection in which we are held by God, who did not spare his own Son, but gave him up for us all, he persuaded many to take up the solitary life. And so, from then on, there were monasteries in the mountains, and the desert was made a city by monks, who left their own people and registered themselves for the citizenship in the heavens.

So testified Athanasius, who had known the saint well.

Long before he had finished that book, before even he had reached the great charges that Antony laid upon his brethren, his interpretation of the monk's vocation, his guidance for avoiding sexual and gluttonous temptations, his longing for martyrdom, Fionán was excited as he had never been excited before. Before this, his faith had been a quiet, a steadfast, a devoted, an obedient thing, but nothing about it had made his heart leap lustfully with an overwhelming urge to belong and to bury his own identity in that of another. Everything about Antony's life attracted him as a lover might: the strength of the man, God's revelation to him, the harmony with all things natural; and the promise of fame, even now fulfilled in this distant land. He sensed that something divinely planned for him had brought him to this monastery, to the tutelage of Brendan, to these inspiring and thrilling commentaries. And, oh, there was something wonderfully *tempting* about it, for it defied all the instructions given you from birth, all the ways of living life here on this wondrously green and gentle and utterly Hibernian earth, where you were watched and protected by, where you shared and contended with, where you submitted and you belonged to a tremendous and wide-spreading community of people, starting with your forefathers and your own family, and including your neighbours, your clan, the local men of god, the more distant chieftains and the great high king, the ard rí himself.

When he put it to Brendan, the old man was not surprised, for in a small way he had planned it: it was at least part of the reason he had taken care to have Fionán priested by the bishop the year before. It was still uncommon for a monk to be ordained. Only four monks out of several score were priests in their community.

'Yes,' he said, 'it is possible that you should follow in Antony's footsteps. And if you should, then nothing can stop you.'

'Even though I would be imitating him, and not the Lord?'

'The Lord went into the wilderness, too.'

'But where can I find such a place here at home?' It was difficult to imagine anything quite as testing as Egypt, in an evergreen land where all things grew fruitfully in dry season, where there was plentiful rain, and where all the rivers always ran full, often to overflowing. And people everywhere, never far away. It would be impossible, surely, to find anywhere remote enough and hard enough to live as Antony did.

'There's nothing on the land that could give you what you want, certainly. But there are many barren rocks around these shores, more even farther afield. They would be as testing as any desert, Fionán.'

'Then that is what I must seek, Abba.'

'But not alone . . .' Fionán opened his mouth to demur. '. . . No, not alone, not in this climate, not on an isolated skeilic where the seas and the storms can destroy everything in their path that is not immoveable rock.'

'But what does it matter if I am destroyed, if I am then with Christ?'

'Not yet. Martyrdom may come one day, all too soon, but remember that it will not be an accident or a matter of choice. Christian martyrdom is always by the will of God. You are not to seek it from the start. You are meant to be a sign before that. You must take companions so that you can make a beginning there at least, so that others may follow on after you.'

The young man lowered his head in obedience, while Brendan, with a pang, remembered his own impetuous and headstrong youth. They had tried very hard to smother his ambition to seek the uttermost parts of the earth.

'Take twelve and imitate the Lord in this also. Now go, for I am tired, and I have my prayers to say.' He was dead before the feast of Patrick came round again.

It was several more years before Fionán and his companions left the monastery to embark on the voyage to find their wilderness. He took his time in choosing who would accompany him, often letting months go by before bringing up the matter with another monk, never approaching anyone who appeared to irritate his brethren by his demeanour, by his habits with food, by his manner of speech, or by anything else. He chose one man because he burst into laughter when anything went amiss as a result of clumsiness or oversight: Fionán was not sure that such levity would have pleased Antony, but he himself had always been attracted to expressions of joy. Not every monk would be suitable for an existence even more confined than this one; but not one of the men he invited failed to give the impression that he would have been badly disappointed had he not been asked.

So many other preparations had to be made meanwhile, beginning with yet more work in the scriptorium, copying the few essential books that they must take with them. Almost the whole of Scripture they had learned by rote, and there were many other things that they had committed to memory. But they were starting something whose future they could not foresee: not where their anchor-hold might be, nor who coming after them might need instruction in the faith. There was, besides, something very comforting, something that promised stability, in these beautiful tomes. Only the Gospels were so large that they filled a man's arms; the other books were small enough to be carried in the hand. But all of them were wonderful to the touch and to gaze upon: their leaves were as smooth as polished bone, their illumination as bright as a heron's eye.

And there was a boat to be built. The river was within sight of the monastery, and on it they used a small curach whenever they needed to fish. Something much bigger was necessary for going to sea, and the monks began to construct it above the river bank. First, though, the hides of eight oxen had to be prepared, being soaked for many days in troughs of water with the ash of seashore plants to make them soft and workable, then scraped with knifes to rid them of hair and fat, before being returned to the troughs for tanning, the water this time being stained with the bark of oaks, in solutions that became progressively stronger until the liquid was so dark that no hide could be seen beneath its surface. Finally, the leather was dressed by anointing it with the tallow of sheep, and polished with smooth stones, so that it should remain supple and be less likely to crack. Meanwhile, ten young alder trees had been felled to provide timber for the framework of the boat, which was made on trestles and upside down. Fionán directed every stage of these tasks, often hurrying to say the offices with scarcely enough time to wash the tannin from his arms first or, later, to tie the half-finished curach down so that no wind could damage it while the monks were at prayer.

Gradually it assumed a recognisable shape, with the long gunwales just above the ground and the ribs springing above in wide arches from one side to the other of the inverted craft, which was without benefit of keel, for it was meant to ride upon and not plough through the waters. For that reason it began with a long rising prow, the better to climb a steep sea, and ended squarely at the stern, with two pegs on the left side to take a steering oar. Strakes were laid along the bottom ribs to give the crew a footing, three benches were set athwart the boat for the oarsmen to sit in pairs, and in the middle was a mounting on which a mast could be stepped. Not a piece of metal was anywhere used in this framework, whose timbers were secured with dowels and joints, or lashed together with leather thongs, recessed carefully into the woodwork so that they would not rub and fray. Only when the frame was complete and still upside down were the hides brought to it and cut to cover the outside of the craft: they were stitched together with cords that had been soaked in the pitch of pine, and were then lashed to the wood, being dampened first so that they would tighten and fit more securely when they had dried. And only then did eight men take the curach in their hands at last and turn its openness to the sky. This was easily done, for the boat was light enough to be blown off course by the gentlest of winds. Brendan had told Fionán that he must beware of that; also that, because the curach had no keel, he must at all costs avoid a sea coming at it from the side, for it would surely be rolled over so, and even the swimmers among them would be lost.

They were, almost, ready for their great enterprise. Fionán was himself now in his middle years, with harsh lines running down either side of his mouth and nose, tiny red filaments beginning to network his cheeks, the long locks of hair flowing back from the half-shaven crown becoming streaked with grey. It struck him, as he saw the boat on the river for the first time—to find out how it sat and test it for leaks—that more than half his life had been spent in preparing for this. Little more remained to be done before they left in the season of harvest, on the feast of all fruitfulness. There was first, however, a fast of forty days to be kept, as Brendan and his company had fasted in imitation of the Lord, before beginning their quest for the Promised Island of Saints. And for that same period, Fionán and his twelve disciples took nothing but a little bread and water every third day after sunset, in order to purify themselves.

It was, thought Fionán, remarkable that the body could be trained over many years to be almost indifferent to food, to retain its strength on much less than the normal measures of fish and bread and plants. But he and the twelve who were to sail with him were glad to be helped by their brethren with the final tasks, when feebleness began to hamper them in the heavier work. There were tools and other implements to be made out of metal and stone, bone and wood, leather and cord and other substances. Provisions were prepared to sustain them as far as their anchor-hold: some bread and some vegetables pickled in brine. Fish they could catch freshly for themselves with either net or hook.

On the day of their departure, the church was crowded not only with the whole community, but also with people from the nearby village, for they too had become involved in this enterprise, hewing the trees and butchering the beasts that had provided the materials for building the boat. And when the mass was done, Fionán went to the altar with some ears of new corn that he had sickled an hour earlier, and placed them in a dish before Christ crucified, just as his ancestors would have done once in honour of Lugh; and the abbot who had succeeded Brendan blessed a vessel of water to make it holy, which was then secured inside the prow of the boat.

The bags of implements and provisions, and the waterskins, were placed carefully out of the way of those who would row, and the bundle of most precious things was stowed in the stern, where Fionán would protect it with his life if need be. Before climbing into their craft, he and his companions embraced their brethren one by one. Although he was anxious to be off, excited by all that lay before them, he was dismayed by the sad weight that lodged itself in him when he stood before his abbot for the last time. They had come into the monastery within a few months of each other, and in other ways had been close, in both affection and temperament, in no way rivals but good brothers in faith.

'Till the next meeting, Fionán,' the older man said. 'God go with us both meanwhile.' Fionán nodded and suddenly dared not speak. Their positions could so easily have been reversed.

The murmuring of farewells ceased as the curach was rowed into the middle of the stream, and the brethren on the banks knelt in a grey disturbance of robes among the great gathering of people who were now watching intently, hands fluttering across their chests to make the sign as the oarsmen leaned into their work. Fionán raised an arm in final salute, and afterwards did not once look back, as his boat began to glide swiftly away behind a long line of osiers growing from the bank, and was finally lost to sight round the first of the river's great bends. It was early afternoon, and the oarsmen were sweating under a cloudless sky, when they began their passage of the lake; the sun was almost ready to lower itself when they had crossed those still waters and entered the river again; a fiery disc beginning to palpitate above the horizon when they came to the estuary and paused to see which way they were intended to go.

Now, on the second day, they went on to the south, each of them wondering at the smoothness of the ocean, such as they had never heard or dreamed of before. Only a scarcely perceptible motion, as in a creature very deeply asleep, which made their craft rise and fall fractionally in lazy swings, told the difference between this shimmering expanse and the waters of the lake. Across the sparkling surface the coast once disappeared, but then emerged again after they had crossed the mouth of a wide bay, when some islands appeared very close to the land. But these were not for them, so near to the habitations of men, and not heaven-sent in their own appointed path. When dusk came before the second night, the curach was allowed to drift again, while long black birds flew low across the water to reach some haven before full darkness fell. Most of them, Fionán noted, were heading away from the land.

'Oremus,' he said quietly, after his men had shipped their oars; and they joined him in the most devoted prayer of all. 'Credo in Deum Patrem omnipotentem . . .'

The words slipped away across the peaceful sea and mingled with the last beacon cries of late-homing birds. The monks broke bread again, and once more huddled in their cloaks.

Fionán woke once during the night, to discover that not only had the moon disappeared but that with it the stars had gone, too. The next time he stirred there was a familiar grey light above them and on every side; but never had he known so thick a fog at this season of the year. Its clamminess reached into the boat with a strangely sour and pungent smell, and it was impossible to detect anything visible beyond the gunwales and the prow. But filtering through it were sounds these monks had never heard before. A screaming such as ten thousand tormented souls might have made was borne to them from some distance away, and Fionán's shocked senses struggled to come to terms with it until he realised that it was the sound of many, many birds. As his heart resumed its normal beat, he also heard from much nearer the noise made by small waves slapping against rock. Suddenly, so suddenly that every man in the curach gasped in astonishment, the fog rose from the surface of the sea in a great swirl, as though some mighty hand had dragged it away, and a beam of sunlight struck warmly where all had been impenetrable a moment before.

They had found their desert. Fionán had no doubt at all about that the moment he recovered from the amazement of it. They were between two immense crags which rose straight from the depths of the sea. The more distant one seemed a little the smaller, it gleamed vividly white in the sun, and it was encircled by clouds of birds cross-cutting the sky just above, while hundreds more were ranged in terraces below. There were not so many birds on the larger rock, but these were near enough for the monks to see their white ordures dripping in great splashes from each ledge. And this was the place they were being drawn towards, the oars still shipped, the monks doing nothing at all but sit and gape at the majesty before them, in awe as slowly they drew nearer still and were more and more sure that this was indeed divinely meant. Cliffs came down to the water's edge so steeply that for long they could not see how or where they might land. But between two surges of the gentle swell the top of a much larger ledge became just visible above the level of the sea. The tide, Fionán realised with another heart's leap, was beginning to ebb: on this ledge they would be able to land and somehow manage from there. He allowed his men to use the oars only across the last small space, to avoid a sharp projection that would have pierced their craft. They tethered the curach to a less dangerous spike, and clambered stiffly onto the wet rock, passing bags, waterskins, bundles, oars, vial of holy water, all that they had, from the boat to their new anchor-hold.

Fionán did not even think about what he did next, for it was as natural to him as saying the Lord's Prayer. He faced the distant land and the revealed power that now hung warmly over it, and raised his arms to the sky in gratitude.

'Glory to thee, thou glorious sun,' he began. His brothers stood around him, hands upturned in obeisance. 'Glory to thee, thou sun,' they chorused. 'Face of the God of Life.' They said it so in the old tongue; and these were the first human voices ever raised in that place, where only birds and seals and the sacred elements had spoken before.

Then the monks turned back to face the great precipice, which reared above them overwhelmingly, invitingly. It was nothing less, thought Fionán, than the very ladder on which they would begin their ascent to heaven.

T W O

AD 670

Bron saw the wall of water approaching him and, in a reflex of sheer terror, fell to his knees and asked for mercy. He had gone down the steep path from the monastery terraces as soon as the wind began to rise, and found it shrieking through the gap between the two peaks like the Ben-Sidhe herself. Before he was halfway down he could see spray shooting upwards in great spouts as the mounting waves began to pound into the far side of their rock. He had laid a net across a cove on the northern side, and it was important to retrieve this before the tempest tore it to bits, whether it had any fish in it or not. Most of the brothers were at the landward end of the skeilic, bringing their curach further up the gully that climbed to the terraces, where it would be more likely to weather the storm once they had got it beyond the reach of any waves and tightly lashed down.

At the gap his foot slipped just as he came from the shelter of the eastern ridge. The wind caught him and would have tossed him far below into the boiling sea if an outcrop had not stood in his way, so that he crashed against it on the very edge of the precipice and lay there bruised and winded for a while before daring to creep further down the steps, which he did crouching low, once or twice on all fours, braced all the while for the sudden gust that might snatch him up again. He paused in the lee of the western peak before continuing down the path, and looked far beyond it across an ominously grey world, slashed with white where the waves crashed and tumbled together but never stopped coming from the uttermost parts of the earth. He could not forget that they were here on the very rim of the world, with an unimaginable void somewhere beyond the limits of the sea; where, he supposed, the earth and the heavens became one in the glory of God.

It was not the waves that were so frightening to Bron, though as he got closer to them they became very frightening indeed. What unnerved him much more was that the very levels of the world seemed to have run amok out there, tilted in planes and with contrary angles that he would not have believed possible, swelling in such vast and swift upheavals that it was like looking at range after range of hills with deep valleys separating each, and all of them moving with unstoppable force towards him. That was what made his mouth go dry and his heart begin to race: the fact that what on that scale was normally solid and safely immoveable, was here liquid and totally out of control; or perhaps very carefully controlled to demonstrate Someone's or Something's omnipotence and wrath. Not only had he never even dreamed of anything like it before, but he found the power causing it so terrible to contemplate that it seriously occurred to him he might just be witnessing the start of another Flood.

One end of the net had been torn loose by the time Bron reached the cove, slithering and stumbling among the rocks above the level of the waves, quickly soaked to the skin as spray poured down on him from high above. The other end could be seen only between incoming waves, which submerged it far deeper than the height of a man. Getting to it seemed quite impossible until Bron noticed that not all of the waves came smashing into the cove: the worst of the sea appeared to be battering at the western peak, whose base offered some protec-

tion where he regularly fished. He balanced himself on the lowest
ledge above the storm's waterline, and waited until the longest inter-
val between the breakers began. He dashed, half-falling, down to the
rope's end, cut the net free with one stroke of the knife, half-gathered
the mesh into his arms, and turned to get back to safety again. He was
almost there when he felt the trailing end of the net catch on something, turned and saw this appalling thing coming straight at him.
From where he stood, it seemed to be almost as high as the peak, itself more a precipice than a wall of water, with green gullies of its
own, an awful curling top to it that was about to collapse, and a foam-
ing, hissing roar that carried above the wind's howl. That was when
Bron fell to his knees and thought rather than shouted, 'Please God,
NO!'; before another instinct told him not to wait for the collapse, but
to try to put even more distance between him and it. He scrambled a
few more paces to some hugely fissured rocks before he fell again, and
this time lay there with his hands thrust desperately into a crack.

He was conscious of the crushing blow across his back, the pain
across his chest, the smothering of his nose and the suffocation of his
lungs, the spinning sensation in his head, the pressing down of his
body, the blackness of everything, the flailing of his arms as he struggled to breathe but choked instead, the utter futility of the struggle, the
end of his strength, and the desire to let it all go; of his body being swept
across stone, and a cracking blow on the head before nothingness
came; and of almost turning inside out as the first gulp of air and not
water was inside him again, then heaving and coughing and vomiting
all at the same time. Voices, which he supposed were angel voices, and
uplifting hands which could only have belonged to Thee . . .

His brothers half-dragged, half-carried him bleeding back up the
cliff, through the fuming spray and the battering wind to the shelter
of the terraces, which rang with quietness amid the storm. They took
him into his cell and, stripping him of his clothes, rubbed him roughly
with wool and placed him between two fleeces, spending a few of
their precious candles to give the room extra warmth. The flames

swayed and bent in the currents of air coming through the doorway, an opening so low that anyone entering needed to stoop deeply. There was no other aperture, and no draughts could penetrate the drystone walls, which were as thick as an arm's length, their stones tightly packed in thin, flat layers. Though the sound of the storm was deafening beyond the terraces, Bon could not even hear it sighing around the cluster of cells whose domes huddled, as if for mutual solace, in two levels just under the ridge. He was glad of the fleece beneath him, for after a night with only dried grasses between him and these slabs, he was generally both stiff and shrammed to the bone except in the lighted half of the year. As he drifted in and out of exhausted sleep, he was also comforted by the glow of the candles on the curve of his ceiling, making it seem much closer, more protective, than when he was without illumination.

He tried, after he came to his senses enough to realise that he had lost a sandal somewhere as well as the net, to imagine what it must be like to be a monk in a land far from this cold wild sea, a country where the earth and rocks were never cool, and where the sun beat down so relentlessly that it was necessary to take refuge from it. He could not envisage this, and he was even forgetting how delightful it was to be warmed by fire and to drink a hot broth to stave off the winter's chill. His mother had been a great one for the winter broth of roots and beans and boiled-up bones, for which wind-dried fish, raw eggs and greens were no replacement except for giving a monk adequate strength. He was no longer sure that the life of the early Fathers in the desert was the most testing, the most purifying sacrifice of himself that a soul could offer. No form of exile, surely, could be more of an oblation than this, even though it was self-imposed and well within sight of the blessed land.

Later, remembering that frightful moment when he thought he would surely die, Bron was not at all clear whether he had asked mercy of the Lord in heaven above, or the more ancient God who lived and reigned upon the waters, as upon (and within) all else; and during the time it took for him to regain his vitality and for his wounded body to heal, he privately tussled with this accusation in the solitude of his cell and when he took his first painful steps again outside, not even mentioning it to his soul-friend, with whom he was supposed to share all matters spiritual. He felt a creeping guilt about that, too, for it was axiomatic that the anamchara could only advise or console or suggest greater discipline if he was privy to everything that entered a monk's heart and mind. There were intended to be no private recesses, no reservations, in such a relationship. Yet any guilt that Bron felt about his secrecy was as nothing compared with the remorse he experienced when he recalled that not for one instant, as he saw that monstrous sea rolling inexorably towards him, had he felt a flicker of exultation at the probability of death, had his soul leaped with joy at the imminent prospect of union with Christ. Instead, he had simply cowered against the rock face, praying—he supposed an incoherent flash of life-wish was a form of prayer, or was that, too, a blasphemy—that on this occasion he might be spared, that he might not be taken until he was properly purified. Abjectly, he recalled that he had not even made the sign of the Cross before the likelihood of death.

It was many days before Bron had recovered sufficiently to resume the regularity and balance of the little community's life: the times when he and his brothers said the offices or in communion made Christ's body together on the leacht in the open air; the times when they withdrew to their cells or to the oratory to meditate or pray alone; the times when each busied himself in the necessary tasks of washing garments, or fishing, or cultivating the small garden, or mending and rebuilding things. The tempest had finally blown itself out by the time he began to move about the terraces again, and even went as far as the gap, though he still shrank from going down the steepness beyond, after very stiffly managing to descend the first few steps. The sea by then was reduced to its grim familiarity, with a yet-powerful but low swell, and waves that still curled and broke from the summit of every one, conditions that a curach might just about survive with a skilled crew, though very uncomfortably. An unbroken heaviness of rain-cloud concealed the western summit of the skeilic in thick mist and completely blotted out the coast, as it had done since well before the storm. It was only just possible to detect the smaller skeilic, and that was close enough for its swirling birds to be visible in good light. Truly, thought Bron with a shiver that was not wholly caused by the cold drizzle, they were at the end of the world.

This was the moment when he wondered for the first time whether he was fit to be there. Had he, son of Bairre the iron-worker, made a dreadful mistake when he begged the abbot to be allowed to join the community? From his own monastery on the lake island, where he had grown from youth to manhood, he had been given permission to go out in the next boat that, twice every year, replenished the skeilic with candles, necessary implements and bread, which was the greatest necessity of all, the lifeblood of the community. If the abbot accepted him on the spot, then it was understood that he would not return.

The old priest, his cloak billowing in the wind, was at the landing to help unload the curach and to take charge of the precious bread, with which he would be empowered to perform the great miracle of transforming it into the very Body and Blood. When Bron knelt and petitioned him, his instinct was gently to turn the visitor away from further formation at his mother house. This pleading snub-nosed face with its rosebud mouth was still without the lines of maturity, its only marks those of anxious youth, though he had already been for several years a professed monk. The chances were that the boy's temperament was extreme, but that he had not yet acquired the absolute control over it which came only after long submission to discipline. The abbot had known men for whom half a lifetime in a mainland monastery was not enough to ready them for such a rigorous regime as the skeilic inflexibly maintained.

'The life is much harder here than on Inisfallen,' he said, 'and it is not meant for every man. Failure here would be particularly cruel.

There can be no going back to a more comfortable life, for the stigma of failure would be impossible to endure in another place: either that or you would become possessed by the Devil himself. The suffering would have to be endured here and it would be more terrible than you can imagine. The forces of darkness would see to that, because they would separate you from God's mercy. You have never looked into the pit, I imagine, but you would find it here if you failed.'

The grey eyes that returned his gaze had not flinched for an instant. There was resolution in them as well as ardour.

'How well are you prepared?'

'As well as I can be, Abba, but I wish to be prepared more. I wish to be tested here. I wish this to be where I shall rise from the dead.'

'Why?'

'Because, long ago, I had a dream in which Our Lord raised me up into heaven from a high place, from the summit of a rock, and the feeling of release I then knew, the immeasurable joy of it, was such as I have never known in any other place, not even at the mass, not even when we have made Christ's body upon the altar. I have had this dream many times since. And this, I think, is the place that was meant.'

'And if despair instead was meant?'

'Christ was in despair when He hung upon the Cross, Abba. But He endured it and ascended into heaven nevertheless.'

The abbot considered him more carefully: it was quite possible for someone to be mature beyond his years. And one of their brothers had died recently, so that a cell was available. Though they did not always succeed in the original intention to order their community in imitation of Christ and His apostles, the abbot and his predecessors had always tried to nourish twelve monks here across the years.

At that moment a fish leapt from the sea behind the curach, every gleaming line of its body, every iridescent pattern of its skin expressing the sheer joy of vitality, the beauty of all life, the boundless possibilities in creation, before it completed its arc and smacked back whence it came, leaving only a trace of bubbles. It was a good and

God-given symbol, whichever way you took it, the abbot thought, and it made up his mind for him against his earlier inclination.

He leant forward, raised the young man from the ground, and blessed him where he stood. 'Then come and be welcome here. I shall hold you forever in my prayers.'

That had been nearly two Easters ago. And now, Bron was at last troubled by the possibility that the abbot's initial reaction should perhaps have been a warning to him that he was asking too much, too soon. That perhaps he was not even spiritually equipped to be any sort of monk, let alone an anchorite.

He sought Assicus some days later, when the older man emerged from his cell, and they climbed up onto the ridge, where they could sit and be alone, with a drop straight down from the knife edge into the sea behind them and, on the other side, the huddle of small buildings just below the steep slope of bare rock at their feet. Bron explained his guilt and what caused it, hugging his arms together inside the sleeves of his garment, tensely rocking back and forth.

'Am I a fraud, anamchara,' he asked, 'have I mistaken my vocation?'

'I don't think so. You're a man who was very badly frightened, as I should have been, too. We are very frail creatures when set beside other parts of creation. Remember Jonah in the belly of the fish. Remember the man of God slain by the lion. Remember all the natural catastrophes visited upon Israel. You were unwise to go for the net alone with that sort of sea running. We could have lost you quite easily. It was only because Abba noticed you were missing that we reached you in time.'

'The net has always been my responsibility.'

'True, but prudence is part of being responsible.' Assicus paused, to reflect for a moment on ways of saying things. He constantly reminded himself that people in his previous life had often found him forbidding on first acquaintance, a natural austerity of manner frequently being mistaken for a low opinion of them. In fact, this concealed a warmth that he kept very carefully controlled, as the abbot had perceived long before asking him to become the newcomer's soul-friend.

'I don't think it is our vocation to be careless with our lives in that sense, to go looking for our deaths,' he continued. 'It certainly is our vocation to be ready for death when it comes to us, and not to reject or deny it. Not to care that it has arrived at last.'

'As I rejected it?'

The older man nodded. 'If you are afraid of death, why do you look forward so much to sleep, which is an imitation of death?' He chuckled fondly, and the long solemn face was transformed by creases and twinkles. 'No one here struggles as hard as you do to stay awake during the vigils, Bron.'

'But from sleep we awaken to the dawn.'

'And from death our awakening is with Christ. You must believe that. Otherwise this life is folly.'

'You believe it?'

Assicus hesitated for a moment, then decided to say it. If Bron was to trust him he had to be truthful. The alternative was a form of treachery. 'Yes,' he replied. 'Most of the time.'

'Not all the time?'

'Not all the time. I'm a weak man, too, you know. And the forces of darkness are very strong. Almost as powerful as the Light, sometimes. But never quite, thank God.' He looked up at a passing gull and whispered into the air. 'Never quite, please God.'

'How do I fight them?' Bron suddenly felt an urgent need to do battle and to be purified by combat; not merely to await something of which he was so unsure. 'Please tell me, anamchara. I do want to walk with God in this.'

'You remember the prayer the Fathers taught? Live it. Breathe it.'

Bron remembered it well, for he had learned it long before he came to the skeilic. It had been part of his preparation for this pilgrimage. He could still recite almost word for word what Cassian had written. That the prayer conveyed all the feelings human nature was capable of, could be adapted to every condition, could be usefully recruited to find every temptation. That it expressed the humility of the most pious confession. That it conveyed the watchfulness which was born of endless worry and fear, a sense of our frailty, the assurance of being heard, the confidence of help always near and present. This one short verse was an indomitable wall for everyone struggling against demons, an impenetrable breastplate and the sturdiest of shields. Whatever our disgust, our anguish, or our gloom, this verse kept us from despairing of our salvation, since it revealed to us Him whom we called, He Who saw our struggles, and Who was never far from those who prayed to Him. If things went well for us, if there was joy in our hearts, this verse warned us not to become proud, because our prosperity could not be retained without the protection of God. The verse was necessary to each of us in all conditions because it acknowledged the need for God's help through good and ill.

All this Bron remembered as well as he recalled any injunction in anything he had ever read. 'Come to my help, O God: Lord, haste thee to help me.'

He intoned the words with feeling, drawing breath in the first part as he had been taught to, letting it out again in the second, falling into the rhythm without thinking about it. He had used the prayer often, though no longer as obsessively as he had when first under instruction, or when he arrived at the skeilic. He had become lax; he realised this now.

'And again and again and again. Let it be the first thing in your head when you wake up, let it be with you throughout the day, let it be on your lips when your slumbers begin so that you are praying it even throughout the hours of your rest. Let it be a continuous prayer, and endless refrain, when you prostrate yourself at the offices, when you rise up to perform all the necessary tasks in the rest of our life here. The moment you find yourself falling away, say it with renewed vigour. If you do this long enough, and devotedly enough, you will one day be purified and worthy of union with Him. Things are beginning to change, Bron, and I'm not sure where we are being led in this life. But this, like God Himself, will never change. I'm much more certain of the life to come than I am of the Church.'

'But how shall I know that I am advancing nearer to Him, that I am becoming purified?'

'You won't, especially if you think of prayer and meditation in that way. You must simply open yourself to God's grace, by excluding everything that might come between you and Him. Open yourself and be still. Nothing in this life is more important than the stillness of it. We are, all of us, on this skeilic because there is a stillness here we have not found anywhere else, and in that stillness we must listen, listen, listen. Be still and listen, my little brother. Open yourself and be still and breathe the prayer. Trust that the Lord will then raise you up. And never again say no to Him. Always say yes, whatever it is. There is no other way.'

Figures were beginning to emerge from the cells below, moving towards the altar at the end of the upper terrace. It was the hour of Christ's death on the Cross, commemorated at the fourth office of every day. Assicus stood and held out a hand to pull Bron to his feet. 'Come, we must go to nóin.'

In three files of four with their abbot standing alone, the hooded figures stationed themselves around the leacht and prostrated themselves, before regaining their feet and moving unhurriedly into the familiar sequences, reciting the words of faith at the appointed place in the liturgy.

'. . . creatorem caeli et terrae. Et in Jesum Christum, filius eius unicum . . .'

The phrases drifted away from the terrace into the mewing of the birds, the booming of the wind, the surging cadences of the sea, until they were indistinct, were finally amalgamated, in a faultless and eternal harmony.

Luke's account of the annunciation was read by a monk from a book that Fionán had scribed with his own hand at Brendan's monastery; and so was the revelation to John of a new heaven and a new earth, but no more sea. Otherwise, the brethren performed their devotions from memory. There was a chanting of psalms, with each file of worshippers voicing alternate lines, running the start of each from the ending of its predecessor without interruption, to produce a measured rhythm that could become as hypnotic in its effect as the rush and recession of waves on a strand. Three psalms were sung, and after each the brethren bowed towards the altar, and when this sequence of chanting was done, all went to their knees except the abbot, who recited the collects for the day in a voice that pronounced these invocations powerfully, as if he expected to move mountains with them. The kneeling figures rose again after they had received his blessing, made the sign and bowed; and returned to their cells to reflect on Christ's release from His agony.

All except Bron, who was restless with the desire for combat still. Letting his hood fall to his shoulders, he made his way down to the cove, to see if the repaired net was still in place: after the storm died down it had been discovered entangled with rocks and badly torn at the far end of the skeilic, and it had taken Bron days to mend it.

'Come to my help, O God: Lord, haste thee to help me. Come to my help, O God: Lord, haste thee to help me . . .'

Dutifully he turned the words over in his head, repeating them in an unbroken soliloquy as he made his descent, adjusting his breaths to coincide not only with the rhythm of the prayer, but also with the tread of his downward steps. At the cove he found the net still secure and drew it in, but it was without fish, so he tied it in place again. He was straightening up from this task when a passing bird dropped a white splat on the bareness of his scalp, just beside the hairline that ran from ear to ear.

'Ach, Gainéad!' he exclaimed in irritation, scooping up a handful of water to wash off the mess. He could feel a stubble under his fingers and made a note to ask for the razor soon. He had long since become reconciled to the tonsure, though it had humiliated him when he first entered Inisfallen and the brother with the clumsiest hands in the monastery had cut his long, golden locks away from half his head. Though he had never seen himself, he could feel his loss, even without

touching his baldness, and he did not much want to resemble his brothers in this respect. Only slaves and monks looked like that, thought Bron
at the time, as he also recalled the Apostles' approval of those who had
made themselves eunuchs for the kingdom of heaven's sake.

Not only had he long since become reconciled to what had at first
felt like another and only slightly less dreadful form of mutilation; he
now wore his tonsure both as a mark of his calling, and as a proud
badge of his wider identity. He was a Celt, a Gael, a Scotus, who not
only followed in the faith of Patrick with all his heart and soul, but
who was deeply conscious that he was also the spiritual heir of the
Tuatha. In childhood, even before his infancy was done, Bron had
learned the ancient stories of his people at the same time as he was
taught the message of the Gospels, and they did not seem at all contradictory to him or to anyone he knew: the wonder of creation was
common to both, they were simply different forms of revelation, and
it was perfectly possible for anyone with half a sense of the mysteries
behind the making of the world to believe equally in Jesus of
Nazareth and in Cú Chulainn. When he played with other boys he
was glad to be the lusty hero of the cattle raid, and his hurling stick
would become the deadly Gae Bolga with its thirty sharpened points,
which broke Fer Dia's shield and crushed his ribs and filled every one
of his joints with barbs: yet he would innocently pray that night with
all the passion at his command, to become more like Christ in gentleness and strength and holiness, and one day, please God, to sit at his
Lord's right hand. His world had always been populated by biblical
prophets and saints and their adversaries on the one hand, by the heroes and heroines and demons of Celtic mythology on the other.
Sometimes the two became one in an imaginative process that Bron
drew freely from without being able to account for it.

His own parents had often mentioned Brigid and Brigantia in the
same breath and saw nothing incongruous in a similarity between the
Christian abbess and the much more distant Celtic patroness of bards.
Just before he left home to enter Inisfallen, he had asked his mother

whether her vision of heaven had ever changed over the years, and the answer was not at all, not in any way whatsoever; indeed, she was more sure than ever that this was the way of it. It was just as the old poem had once described it, an island of the good with gently sloping hills of green, under transparent clouds which nurtured streams that made pleasing sounds, like the faint notes of a half-touched harp. All was calm and bright there, with the pure sun of autumn shining from his blue sky upon the fields; and upon the rising hills were the halls of the departed, with high-roofed dwellings for the heroes of old. That sun was the Son of God, Jesus Christ our Lord. And she hoped to see Him with her own eyes before long. Her vision of the eternal was the one Patrick had brought to this land, grafted onto the vision her people had always held when they pondered the significance of the world into which they had been born.

Bron fingered his scalp again and frowned as he remembered something Assicus had said. There had been rumours for some time now that the Church in Britannia had been placed under pressure from Rome to conform to its practices or be declared an heretical sect. There had been talk of the Celtic tonsure no longer being acceptable, of the monks in Dalriada and Northumbria and elsewhere across the water being obliged to adopt the Roman manner of shaving the head and marking the monastic vocation. Some even said that the Pappa of Rome, sitting on his cathedra beside the Tiber, had demanded that the Celts abandon their traditional time for celebrating Easter, which they had received from Patrick, its chronology deeply rooted in the Hebrew books as well as in the Gospels, in the established method of determining the Jewish Passover. The Celts were being told that the most important day in the entire Christian calendar must now be altered on the say-so of a Gallic bishop whose alternative calculations, dubious as they might be, had without hesitation been adopted by the bishops of Rome. The incompatibility between this and the Celtic observance, it seemed, could in some years be as much as a month or more.

Their own faith in the Celtic Church had never come to them through Rome, but more circuitously and very probably by sea. But it had every bit as much validity, for its source was also in the Gospels and in the life of the earliest Fathers of the Church. There was no reason at all why they should submit to such pressure, especially to pressure coming from the Tiber's banks. Romans, Bron was taught early in life, had always been the enemy of Celts, had slain their heroes and heroines wherever these stood firm in the aggressor's way; and through greater military might the Romans had driven the Celts from one country to the next, cruelly, remorselessly, inflexibly, until the Celts had nowhere to go but this very edge of the world. He wondered if it might be possible to fight this new Roman domination in the Church; whether Assicus and Abba and his other brothers were so inclined, too. Things might be changing not only in Britannia and Dalriada, but even on the mainland opposite: out here, however, they didn't have to change too much. No bishop had any authority here, certainly no episcopus from Rome.

Guiltily, Bron realised that he had not been saying, living, breathing the prayer. His mind had been anything but still as he stood there and frowned at the sea which, he now noticed, had lightened since he came down to the cove. The weather was beginning to lift at last. And he must make reparation to Almighty God, on his knees in the oratory at the other end of the skeilic, without delay. He started back up the path, and when he reached the terrace he saw that the land was just visible for the first time in many days. He went into the small chapel, with the long ridged roof that gave it the shape of an upturned boat. He knelt on the stone floor and began to say the prayer out loud, carefully measuring the words with a deliberately unhurried breath.

He had been there, entranced, scarcely aware of the passage of time until he realised that the numbness in his legs was now high above his knees. You could always tell how long you had been praying by how far an ache had climbed up your limbs, and by how much they had subsequently gone dead. For the first time he looked up, and his eyes came level with the small opening that had been slit into the end wall. Through it he could see the distant hills quite clearly now. And, much nearer, coming closer all the time, a curach under sail, rolling and vaulting over the sea, bringing its cargo of necessities and news. Bron drew breath once more, and forgot the discomfort from the oratory floor.

'Come to my help, O God: Lord, haste thee to help me. Come to my help, O God: Lord, haste thee to help me. Come to my help, O God: Lord, haste thee to help me . . .'

THREE

AD 780

The harshest days of winter had passed, and with them the worst of the storms. But almost any month of the year was liable to produce seas that came rolling into the skeilic with such force that its lower half was made treacherous by wave or by spray; and, when this happened, the tremendous rise and fall of the swell against the outcrops and buttresses meant that no curach could possibly leave or land on the rock. Similarly, there was no week in the calendar that could be guaranteed against driving rain and dense cloud, which might blot out the sight of the western peak from the monastic terraces for an entire day, or make it reappear periodically, fantastically, as if belonging to another world, from behind the smoking banks of mist. Without such ample rainfall the skeilic would have been uninhabitable, for it had no springs; but two cisterns had been chiselled out of the ridge, with slanting grooves cut above them to channel the water running down its mighty slabs.

When the lighted part of the year began, the truly overwhelming tempests, which sometimes seemed to threaten the existence of the skeilic itself, no longer raged in this place. The oppressive gloom of winter was also lifted, and with its departure the spirits of even the most dogged anchorite rose, glad that the most biting cold had at last relented, too. Birds, which had been missing through the darkened time, returned to nest again on the ledges they had abandoned months before, and for a little while the community could look forward to their eggs. But their clacking chatter, their screeching din, was sometimes so loud that the monks could scarcely hear themselves chant, and it obliterated any sound that might have carried across the water from the larger colony of birds on the neighbouring skeilic.

So many of these were now filling the sky around the lesser rock each day, or roosting upon its terraces, that an offshore breeze brought the rank odour of their droppings as far as the monastic garden, where Cainnech was repairing the wall that prevented the thin layer of soil from sliding straight off the slope of the cliff into the waters below. He wrinkled his nose in distaste as he inhaled the faint but unmistakeable sharp and musty smell, and paused after carefully placing one slab of stone on another. This was hazardous work, because if he failed to balance them, if he hadn't dressed them properly so that they fitted snugly against each other, the wall might very well collapse into the sea, and take him with it as well as half the garden. It was a long way down to the sea, even though the garden was some distance below the ridge. There were more than six hundred steps from the base of the rock to the monastic terraces, and from where Cainnech crouched to peer over the wall, it was only just possible to decide which kind of birds those were, swimming and diving directly underneath him. One of them surfaced and then rose clumsily from the water, but became immediately graceful the moment it was entirely in its natural element. All God's other creatures, he thought, were just so in whatever state it had pleased Him to situate them. Only fallen man could be ugly in his habitation of this world. This, to Cainnech, was a powerful proof not only of heaven's existence but of its purpose as well, proof that every form of human grace would be achieved there, and that this world was merely a place where man might be prepared for everlasting glory above.

He followed the long black broigheall's progress towards the little skeilic which, he saw, was beginning to pale again now that its population had returned, for in their absence the spray and the rain of winter had cleansed it of their filth. One of creation's mysteries that Cainnech and his brethren contemplated at least twice every year was where the birds went to when almost all of them disappeared after their young had become strong enough to fend for themselves; and whence they came home to rear fresh broods. The puifin, he knew, tunnelled to make a nesting place wherever there was soft earth, and he was strongly of the opinion that all birds must do likewise in winter, going into hibernation like the hedgehog and the snail. The mystery was that they did not do it here, in their island home, or anywhere else that any of the monks was aware of. The puifin vanished before summer was properly finished, and only the gainéad was still here when Samhain arrived, and with it the beginning of the ancient calendar. Both were now making nests and choosing mates, the puifin's startled face beginning to change from its winter darkness to the hue of a yoke; and presently there would indeed be eggs. Cainnech's stomach murmured with pleasure at the thought, though the days were still distant when the monks could enjoy this delicacy, for the greatest of the fasts, the Lent of Christ, was only half spent and it would be past Easter before the eggs were laid. The time when the monks had eggs a-plenty to vary their eating was very limited, for the blessed things soon became inedible, and seabirds rarely had more than one brood. Cainnech smiled as he remembered a wry saying that had always made light of troubles in his family. 'When the good Lord closes one door, he generally shuts another.'

His attention was diverted to a shelf of rock a little way beneath his wall, where a gull had just mounted his hen and was balanced on her squatting back, wriggling ecstatically to be inside her when she obligingly raised her tail. He watched, fascinated and guilty, but with not enough shame to turn away until the cock had done, and had stepped carefully off his mate and begun preening himself, while she cackled with satisfaction, and rearranged her wings. Cainnech still found the coupling of birds disturbing after a dozen Easters on the skeilic and many more as a monk but, unlike some of his brothers, he had never disciplined himself to look away either out of delicacy or from a fear of being atavistically roused. He took pleasure in every part of creation, in each evidence of life, on this balmy day when the campion which carpeted all but the barren parts of their rock was beginning to bud, and when the greens in their garden had started to grow new shoots. The sun shone warmly upon the gentle swelling of the sea, which rose and fell softly, thrillingly, like the breathing of a maiden's breast. Cainnech stirred at the image, and was uneasy with himself. Quickly he placed the remaining stones on the wall and turned to go and fetch more. He climbed to the upper terrace and then went along the path that led out of the enclosure, striding swiftly, energetically, to drive the unthinkable away.

When he reached the gap and the saddle between the two heights, Macet was nowhere to be seen. They were supposed to be walling the garden together, hammering and splitting the stones here, where boulders and smaller rubble abounded, then carrying them back up the steps to the enclosure and beyond. This was the way all the building had been done on the island from the start, and Cainnech had never ceased to wonder at the devoted tenacity of Fionán and his other predecessors, who had so laboriously created their habitation and their anchorhold out of the bleak skeilic itself, without any assistance, without any means other than their own inspired energy and a few handy tools.

Their first task would have been to look for fresh water and, finding none, to chisel those runnels from the rock. But what certainty they must have had to stay here in the face of such an obstacle, what tremendous faith that they could and would survive here in this wild and empty place. Had they never for one moment been tempted instead to sail away from the skeilic, to seek an alternative which at least offered water to drink? And he had often wondered what shift they had made for shelter until the cells were built, and how long it had

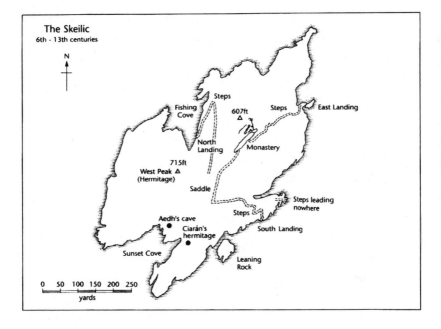

The Skeilic
6th - 13th centuries

N

Steps
Fishing Cove
607ft △
Steps East Landing
North Landing
Monastery
715ft
West Peak △ (Hermitage)
Saddle
Steps leading nowhere
Steps
Aedh's cave ●
Ciarán's hermitage ●
South Landing
Sunset Cove
Leaning Rock

0 50 100 150 200 250
yards

taken them to create the great staircase that rose from the level of the sea and wound its way up the face of the cliff, almost to the top. Some steps had been cut out of the living rock, but most of them were meticulously constructed out of dressed blocks, wedged into place with smaller stones. They were so heavy that few could have been put into position by a single man. He certainly had never seen anything like it in his life, and he seriously doubted whether such a thing existed anywhere else. He did not believe any construction on this earth could possibly be better founded than this. That it should have been achieved in this remote place, in these conditions on the edge of the world, was surely nothing less than a miracle. Cainnech felt his own responsibility in this matter keenly. His heart swelled with gladness when, busy with mending the walls or replacing loose stones on the stairs, he reminded himself that every bit of this was a sign meant to last through all ages; and that he, for the time being, was its custodian, charged with ensuring that this destiny would be reached.

He shook his head and fetched a sigh when he realised that Macet had abandoned work for the day yet again. Cainnech had left his brother applying himself after a fashion to a boulder, tapping it experimentally to identify its weakest seam, as though he did not really expect or wish to find such a thing. He had that dreamy look on his face which said he would prefer to be purifying himself in some way that did not produce blisters or bruised fingertips. Macet had very slender fingers, which caressed everything he touched, as if his hands had been delicately bred to hold a quill or a brush in a scriptorium rather than be clenched round a hammer and chisel; unlike Cainnech's rough talons, which had been gnarled almost from youth and, when clasped firmly together in prayer, conveyed the simple piety of the swineherd or the mason, rather than the spiritual obsession of a man who had cut himself off from the world in order to be nearer, and more certainly closer, to God. Macet always managed to handle tools with an air that suggested he was not quite sure what each of them was for, was even a little intimidated by them, anxious lest they might do him some injury. A walling stone had once slipped a hand's breadth, no distance at all, onto his foot and his yelp of anguish had brought two of his brethren running in alarm. Cainnech would have merely gasped at the pain, and might very well have subdued it silently without even a change of expression across his thick-set face.

This was not the first time that Macet had been found wanting when the two laboured together, and Cainnech had generally tolerated the other's negligence because he liked his smiling gentleness, which had often lightened the austerities of an otherwise gloomy day. And unless the work was heavy Macet was often enough a willing helper; in summer it was invariably he who, without being asked, brought a horn of water for the refreshment of his companion, because he had noticed that Cainnech was becoming parched. Therefore only once had Cainnech demurred when he found himself completing some laborious task in solitude, eventually having to carry back to the terrace the last load of stones and the tools that both of them had used. Macet had not even apologised, but instead he pleaded his greater need to meditate without distraction, the priority he had to give to refining his soul in the service of the Lord, over artisan work that others could do much better than he. This so dumbfounded Cainnech that, sensing imminent conflict between them, he retreated, blinking in bewilderment. Either Macet was so arrogant that he believed he was operating spiritually at a higher level than the rest of them, and could therefore ignore his communal responsibilities; or he seriously misunderstood the spiritual value of manual labour as a defence against accidie, the weariness of heart that visited every monk. Cainnech was not sure what was the cause of his brother's indolence, and was loth to mention the matter to Abba. But although he prayed fervently each day that Macet might be pointed anew in God's holy ways, that they might become true brothers in Christ, the lack of contrition and the fact that nothing changed increasingly rankled with him.

It was a little while before he noticed the errant lying, face down in shadow, where the saddle swept up to the western peak. Exasperated, he strode over to the prostrate figure, which gave no sign of having heard his approach.

'Are you ill, brother?' Cainnech asked severely. There was no reply, and Macet did not even stir.

'I asked you in all charity whether you had a fever.' A threateningly harder edge was in his voice now. 'Or have you simply surrendered yourself yet again to your habitual sloth?' That was unjust, for Macet notoriously slept less than any of them, often climbing down from a lonely vigil on the ridge when his brethren emerged wearily from their cells to say Lauds and Matins in the middle of the night. It was deliberately unfair, for Cainnech intended this to hurt.

It reached his mark. With a groan Macet turned over and rested his supple frame on an elbow, shading his eyes with the other hand while he looked up into his tormentor's face.

'I should have thought,' he said, 'that someone whose snores can generally be heard three clocháin away, often without the decent excuse of darkness, is not the one who should be speaking to me of sloth.'

Cainnech went scarlet and opened his mouth, but Macet carried on before any sound could get out. 'I've told you already, brother. Each of us has his distinct and individual talents. You are the best waller we have. Everyone acknowledges that; it's why Abba has asked you to oversee all such work. As you yourself are so fond of saying'—he wickedly mimicked one of Cainnech's more plodding pronouncements—' "The Desert Fathers wove mats so that their hands should be busy, but we build things that last because we keep them in repair." But I don't have your dedication to the simple crafts, brother. I am likely to make mistakes that you would then have to unmake. I already have done, more than once. Don't you see'—he smiled pleasantly but this, too, was to taunt—'that I'm much better employed praying for your immortal soul; as I do, every day of my life, I promise you.'

'How dare you speak . . .'

'To you like this? I do dare Cainnech, because I am authorised by Scripture to say these things. Our Lord Himself, if you recall, chastised Martha, whose particular talent was very much like yours, when she complained about her sister, as you do about your brother, because she was sunk in the mire of jealousy and self-pity.'

The monstrous conceit behind this parallel almost robbed Cainnech of the ability to speak further; but at last he spluttered his retort. 'Well, I'm tired of playing Martha to your Mary, Macet. I have a soul that needs purifying, too, I have a devotion to the Cross that I would spend on my knees, without another thought in my head. If that,' he added bitterly, 'is truly what you are about.' He wished straightaway that he had not let it out, but he was too far gone in anger now to stop. He would repent alone afterwards, seek forgiveness of Him, not him.

The self-satisfied smirk had left Macet's face, and coldly open hostility had taken its place. Cainnech pressed on before his brother could reply.

'I tell you now that if you once more walk away from your work before I say enough has been done for the day, then I shall denounce you to Abba as someone who is unworthy to be a member of this brotherhood.'

This was playing with fire, as he well knew. Anyone who took such an allegation to the abbot would be subjected to the most searching inquiry, to determine whether the grievance had been expressed in all charity with the intention of helping a brother to purify himself, or whether it was founded in malice. If the abbot decided a monk was malicious, he would certainly exact the most rigorous penalties. Men had been packed off into exile for not much more. Columcille himself was such a one, but few would expect God's mercy to intercede on their behalf, as it had done for him.

And so they went on, mauling each other with words, until the bell rang for the office, and Cainnech turned away at last.

'We must go to Vespers. You can fetch that tackle over there and bring it back.' His companion sat up with tightly clenched face, and buried his head in his hands until the other had gone. Cainnech strode away grimly, his sturdy shape hunched with determination, but sick to its heart with the violence that was there.

Macet followed slowly, the tools slung over his shoulder, his head still bowed and his eyes on every step he took, the muscles flicking along the length of his thin, pointed jaw. He went to his place at the office directly opposite Cainnech, and no one could have guessed from their intonations or their movements that so much anger had passed between them only moments earlier. At the chanting of the psalms, Macet's high tenor slipped perfectly behind Cainnech's deeper tone in the joining of each line, and the same tender conjunction was made when the other's turn came. At the Credo they were utterly in tune, with nothing to rouse the slightest suspicion that their voices had just been raised in contempt against each other, or that they did not see their vocations in exactly the same way.

'... Dominum nostrum; Qui conceptus est de Spiritu sancto, natus est Maria Virgine ...'

And when the office was over, after the ritual prostrations to the altar had been made, they bowed to each other as deeply, as gravely, as courteously, as every other monk in the three files did in bidding farewell to his brethren opposite. Then went to their cells, faces hidden within their hoods, hands concealed inside their sleeves, their differences made as anonymous as their lives, and they seeming as untouched by animosity as if this had been the longest farewell, with them utterly reconciled.

For days their lives coincided only at the offices, when they conducted themselves in exactly the same way. The weather had turned wet again and so there was no call for them to be working alone in each other's company. As usual, each monk on the skeilic ate his daily meal in solitude, after the brother whose turn it was to attend to their victuals that week had placed the communal food on a ledge beside the cisterns just before the setting of the sun; one large wooden platter bearing morsels of fish, another laden with leaves of spinach and stalks of celery. Often enough, some of this bounty was left unconsumed, because one or other of the brethren was resolved to mortify himself even more by eating less and less, the better to purify his soul, even though every week outside the great lenten periods already contained two obligatory days of fasting. Macet was such a one, and though he dutifully went to the platters when the bell announced their readiness, their contents were little reduced by his appetite. And Cainnech made sure that his own movements along the terrace did not coincide with the other's. On two occasions he was about to emerge from his cell when Macet passed by, which caused him to retreat into the shadows until his antagonist had returned whence he came.

It was Macet who could endure the breach between them no longer. After five sullen days of their avoiding each other, he resolved to remedy matters with heartfelt apology and full explanation, trusting that Cainnech would respond with similar generosity. For Macet recognised a genial nature beneath the temper, in spite of Cainnech's inability to see that the exertions he rejoiced in and performed with little apparent fatigue, very quickly weakened his brother, frequently to the point of collapse. He was quite unaware that on the day of their quarrel, Macet had swooned where he worked at the boulder and, on reviving, had crawled on hands and knees to a place where he could gradually recover his strength in the shade. That was not the first time this had happened, and he ought to have mentioned it before. It occurred to him that what needed subduing even more than his flesh was his pride, for that could lead to damnation as well.

He made his move after Terce, quickening his stride to catch up with Cainnech when the office was done, and plucking at the other's sleeve before he could enter his dwelling. Cainnech turned with a wary look in his eyes, obvious even though they were sunk well inside his cowl. Speech was discouraged near the cells, for fear of disturbing their occupants, and Macet simply gestured with an eloquent forefinger that they should move out of the enclosure, where they could be alone. He led the way, turned once to make sure that he was being followed and, after passing through the opening in their high boundary wall, carried on down the first cascade of steps until he reached the saddle. Not even there did he stop immediately, but turned off to the side, above the unpaved path to the northern landing, pausing only when a buttress of rock concealed him from any casual observer. When Cainnech arrived, Macet was waiting for him with his hood down, and an expression on his face that was both sad and embarrassed. Cainnech allowed his own cowl to fall, as he broke the long silence between them.

'Well, what is it?' he asked, cautiously. There was little warmth in his voice.

'I want to apologise and I need to explain something, brother.'

'Go on, then.' Cainnech nodded briskly. Repentance should never be easy, and only accepted if it was made in full, without the slightest reservation.

'I spoke harshly, unjustly, with a desire to wound, and for that and the hurt it must have caused you I am truly sorry. I beg forgiveness if you will be merciful, brother.'

A small movement of the other's head. Nothing more. Cainnech was at least prepared to hear him out. Macet took a deep breath and continued with the harder part.

'I am much weaker than you know, Cainnech. Weaker in body, weaker in spirit, weakest of all in resolve to do my part in spite of my infirmity. I wish I could stand alongside you at all times, work the stone like you, move mountains like you, be tireless like you, do more than my share in the vineyard like you. But I can't. I haven't your strength. I'm as feeble as a woman compared with you. I wish I weren't but that's the way I am, and I do try my best to overcome it.'

Cainnech realised with a sense of shock that Macet was close to tears. He had not known such a thing before, except in women and children. Macet began to tell how he often fainted during great exertions, how he was so ashamed of this feebleness that he preferred to imitate an injured animal and be alone to lick his hurt rather than cry for help from his brethren. He was so clearly in distress that Cainnech's indignation drained away, and in its place came a compassion that he had never experienced with this one before.

'You know what the trouble is, Macet? You're not eating enough. For God's sake, man, this is Lent and hard enough for any of us to bear. But I've watched how little you take from the platters when we're allowed to break the fast, and it isn't enough to keep body and soul together in a life of prayer, let along hard labour on top of it.' Cainnech was always the practical one; and something had put the warmth back in his voice.

'But Antony . . .'

Cainnech waved the comparison away. 'Yes, I know Antony is supposed never to have eaten more than every few days, and there's the legend that Simeon the Stylite starved himself completely from beginning to end of the great Lent. Personally, I've always doubted the accuracy of some of these tales. But even if it were true, Antony wasn't rebuilding walls like us: weaving mats is nothing, compared with what we do. The Stylite wasn't doing anything at all but sit and meditate. And they weren't living in a climate like ours. Never forget that. You need fuel to keep the fire blazing within, for this place will surely put it out, otherwise. Basil is a much better guide than Antony in this respect, Macet. Temperance and discipline is the thing, not starvation.' He pulled a face, and it was amused, wry, above all affectionate. He might have been mildly reproving a son after some incautious prank.

Macet nodded, only half-convinced, but relieved that Cainnech's voice was no longer hard. He was approachable again, his trusted friend once more, and they must seal their amity now without delay, with ties that belonged only to them, so that it could not easily be broken a second time.

'I want to be truly your brother,' he said. 'We cannot let this happen to us again. We have always been bonded in faith. Let us be close in affection, too.' He watched Cainnech very carefully as he began to remove his outer garment, then unfastened his tunic and exposed his breast.

'The old way?' He held out a long, spindly arm and studied the other's face.

For a moment Cainnech stood transfixed, appalled by the other's nakedness. None of us, he thought, has any flesh to spare but this was nothing but skin and bone; you could count every rib down that side, every sinew in the arm. They were not just visible, they were straining against the skin that covered them. It was appalling, and yet it was very beautiful, a body that had been stripped of all grossness, become utterly refined. Most beautiful of all was the poor taut breast, with its veins standing out, with its straggle of hairs and the nipple in its circle of brown. Christ's body would have been like that, at the end, before they crucified it.

He knew what he must and would do: badly wanted to. Those tender fingers had enclosed the back of his head and were slowly drawing him in. His faith and his ancestry tumbled together in confusion round his head, making him dizzy with their turmoil. So much was contradictory, so much else rang true. 'This is My Body which is given for you . . . In that sweet country I could rest my weapon . . .'

Cainnech allowed the hand to guide him slowly to the nipple, opened his mouth with reverence and gently sucked, felt the salt taste on his tongue, Macet's body stir under his hand, was aware of his own beginning to flex. '. . . I could rest my weapon there . . . Lord let me never be confounded . . .'

'What are you DOING!'

The bodies sprang apart as if a thunderbolt had come between them, and Cainnech caught a glimpse of Macet's horrified face before he turned to see the abbot a few paces away, standing like a pillar of wrath where the staircase came down to the saddle. The old priest took a couple of steps toward them then stopped again, shaking with incredulity and rage, while they returned his glittering stare open-mouthed with shock.

'I ask you again. What wickedness is this? Are you bedevilled, the pair of you?' He sounded as though he was choking on his anger.

Macet, trying to rearrange his clothes, could feel waves of nausea threatening to engulf him. 'We had quarrelled, Abba,' he stammered. 'And were now reconciled and . . . and . . .'

Cainnech was swift to his aid. 'It is the tradition, Abba,' he said. 'Our forefathers always did it so.'

'And our forefather Patrick said that the mouth which lies, murders the soul. I say you would do well to remember that.'

'We speak the truth, Abba,' Cainnech said. 'My lips were not be-fouled by what I did. I touched my brother only out of charity and as a sign that henceforth we should both make amends. It is the old, old way of healing such wounds. You must know that.'

The abbot knew it well, was perfectly aware that the old rite was in-nocent of carnal lust—to start with at least. He still disapproved of it, and not wholly because there was no telling where such heedless in-nocence might lead. He was as much a Celt as any of his monks, but he was also a new breed of father in God who, long before coming to

the skeilic, had embraced the changes in the Church that had been ordained by Rome, and which some had fought bitterly until they were
isolated and obliged to submit, rather than stand firm on their native
instincts and accept the living death of being excommunicated for
heresy. Even Columcille's Ioua, the most obdurate of all Celtic out-
posts, had long since, at long last, accepted the new order, the new
tonsure, the new celebration of Easter, the Latinising of the office
hours and all the other changes that had been enacted across Europe
by papal decree. The Romanisation of the faith was complete, the
Celtic Church was no more as a distinctly separate entity, and the
abbot for one was glad of it. God alone knew what perils the Church
might face in future, what persecutions lay in store for it. It would
need all the strength it could gather unto itself, and that came best
through unity.

Although he was glad to be a Celt rather than of any other breed,
he was impatient with those who dwelt in the tribal past, who measured everything by what their forefathers had done, who believed in
the old legends as fervently as they did the Gospel of Christ. The local
heritage included a warrior instinct, which had still not been purged
from the land by the Messenger of Peace. Dear God, even monasteries here had been known to do battle with each other, with corpses
strewn afterwards across fields of blood. The abbot was never reluctant to point these things out to his brethren, reminding them in particular what some who came to him appeared not to believe: that
Patrick himself, when little more than a child, had been abducted
from Britannia by Niall Naoi Ghiallach, whose fleet of curachs had
gone raiding there for loot and slaves.

No, he was not much a believer in the old ways. He saw too clearly
that they could be made the excuse for infidelity to this life, that they
could corrupt where they appeared to be most benign. As in the case
now before him. He was well aware that some rift had opened up between the two monks, and he cursed his own fault of omitting to
speak paternally to both, some days earlier. He had noticed Macet's

gesture to Cainnech after Terce, had been delayed a moment or two after it; but he now blessed the instinct that had caused him to follow the brothers and come upon them before they were hopelessly enmeshed in sin.

'I do know it,' he said coldly, 'and it has no place here, for it is a pagan, a dangerous thing. It must never happen again, on peril of your souls. No particular friendships so signified can be tolerated in such a community as ours. Of that you must repent, both of you, before your brothers. You will go up to the garden terrace and there you will each say alternate lines of the Beati, crosfigel, for the rest of this day. When you come to the end, you will start again, and you will do this repeatedly, until I tell you to stop. You will say it in the old language, and if your tongue slips you will fall upon your knees and not rise from them until you have come to the end of the psalm and are ready to start again. But remember, all of it crosfigel. Now go, and begin as soon as I reach the leacht.' He stepped aside to let them by, their bodies deeply bent as they passed.

It could be much worse, thought Cainnech, stumbling uphill ahead of the others, his head still in a whirl. He could have extended our lenten fast indefinitely, or ordered us to flog ourselves; he could have made us sleep in one of the pools for any number of low tides, or upon a bed of sharp stones for the rest of this year; he could have done almost anything to chastise us, for the Church had an inexhaustible imagination when it came to devising punishment. There had once been a penance here before his time, which required steps to be cut down the steepest part of the cliff, beginning high above sea level and ending nowhere. To teach the futility of a disobedient life, so it was said, and that would have been a hard one to bear: dangerous, too. What a meditation must have gone on there while the penitent worked. Most penitential acts were meticulously codified, and well known to every monk in the land. Some of them had to be performed and endured for months, not a single day. But repeatedly saying the Beati crosfigel would be bad enough, for it was by far the longest of all the psalms, and by the end their arms would be ready to fall from their bodies. Making them say it in the old language was an especially ingenious way of punishing their particular fault. The psalms were familiar to them, always chanted by them, only in Latin, not in their native tongue. They would have to translate as they went along. More opportunity for slips, and crosfigel on their knees. Cainnech hoped his brother would be able to get through it all without collapsing. He had never heard of much mercy being shown in the exaction of penances.

They stood side by side, with room for another between their extended arms, and Cainnech began the long abasement.

'Blessed are the undefiled in the way:'—he paused as he would have done with the Latin text—'who walk in the law of the Lord.'

Before his lips closed on the last word, Macet opened his own on the first syllable of the second line.

'Blessed are they that keep his testimonies:'—their pace would be as exactly measured as if they were in choir, until fatigue began to tell—'and that seek him with the whole heart.'

The phrases flowed steadily and the voices had never sounded more confident, as they recited the first entirety of the psalm, decade after decade after decade of lines, more than eightscore and ten in all. The arms never moved, and the two heads remained proud, not humble, as they besought Almighty God to be merciful unto them, to remove them from reproach and contempt, to quicken them in his righteousness, to leave them not to their oppressors, to give them understanding that they might know his testimonies, to deliver them according to his word; and as they promised that they would meditate upon his precepts, that his testimonies were their delight and their counsel, that they would run the way of his commandments, that the law of his mouth was better unto them than thousands of gold and silver, that he was their hiding place and their shield, that their flesh trembled for fear of him and that they were afraid of his judgements, that they rejoiced at his word, as those that had found great spoil . . .

'. . . I have gone astray like a lost sheep: O seek thy servant, for I do not forget thy commandments.'

Macet was hoarse as he reached the last words. Cainnech could feel the gum starting to form in the corners of his mouth. They began again, only a little less confidently, and had almost finished a second time when the bell rang for Sext and, on the terrace above theirs, the rest of the community came from the clocháin and their tasks to gather at the leacht. The two penitents fell silent as Abba called his other brethren to prayer, and they remained so, arms still outstretched, until the office was done. Then went on as before, as their brothers dispersed.

It was on the third recitation that Macet uttered two Latin words by mistake and dutifully went to his knees to complete the psalm. He was stiff in going down and, before they began the fourth round, he came back to his feet like an old man, with a grunt of pain.

'. . . Open thou mine eyes: that I may behold wondrous things out of thy law . . .'

By the time None was finished, the arms were no longer horizontal, but drooping and well on the other side of pain. Cainnech could feel his shoulders locked in agony, but beyond them there was nothing left. Macet's head was hanging almost to his chest, and there was a white froth upon his lips. Another monk, going down the path after the office to tend the fishing net, looked back over his shoulder, saw the two figures stark against the eastern sky, and remembered Golgotha. He did not pity them. Lucky they were in their small imitation of Christ.

'. . . Let thy mercies come also unto me O Lord: even thy salvation according to thy word . . .'

Five times they had said the Beati, and both had obediently gone to their knees more than once after faults. They were beginning another round when Macet stammered so badly that he could not finish his line. With a groan he sank to his knees, which gave way under him and he fell to the ground. He lay there, crying softly in humiliation and pain, tried to rise, but his limbs would not respond. Cainnech, looking sideways over a sloping arm, thought Macet was almost done.

His own throat was on fire and simple breathing was hurting now. His eyes were stinging and blurred so that he could no longer see clearly beyond the terrace to the white-haired sea in the south.

Footsteps across stone, then across the ground behind them. Abba placed a bowl of water on the wall then bent his mouth to Macet's ear. 'We fall down and we get up in this life. Always that. It is a great truth.' He reached for the bowl, held it to the younger man's mouth. 'Come, my son. Drink this, then rise and continue as before.' He did not help Macet to his feet. Nor did he offer Cainnech what Macet had carefully left in the bowl. He threw the rest of the water over the wall but the wind blew back some drops, which slashed Cainnech like a whip.

'. . . Trouble and anguish have taken hold on me: yet thy commandments are my delight . . .'

The light was beginning to drain from the sky, and Vespers had passed. They were coming to the end of the psalm for the seventh time and Cainnech's voice was almost incoherently dry, Macet's no more than a whisper, the rhythm of his lines gone, his pauses sprawling across words he no longer controlled.

'. . . Princes have persecuted me without cause: et a verbis tuis formidavit cor meum . . .'

He went down with a crash, and as his head struck the ground he knew that he was descending into hell. He had lost his Lord and he stood alone on the edge of an enormous pit from which great flames rose high into the air before falling back again. He saw that they flung up the souls of men like sparks in the belching smoke, before these, too, dropped back into the abyss, from which came the stench of burning flesh and disgusting filth. He stood there terrified, and heard behind him the sound of great weeping, and with it came loud laughter as though an army was insulting and enjoying the misery of captured foes. He turned and saw a crowd of evil spirits dragging the souls of many men, which were wailing and shrieking in the thick darkness surrounding the pit. The evil spirits and the souls plunged over the edge together, and as they went deeper Macet could no longer tell the difference between the evil laughter and the terrified shrieks. But then there appeared behind him the brightness of a star, in which he saw the Lord returning to him with hands outstretched. He led Macet towards the light, which revealed a boundless wall, with no end that he could see to its height or its breadth. Yet the Lord took him to its summit, from which he beheld a wide and pleasant plain full of flowers, and a multitude of men and women in white clothes, all bathed in a light much greater than the light of day. He heard sweet singing and the air was filled with a fragrance such as he had never known before. He wished to join the multitude in this beautiful place, but the Lord led him back by the way they had come together. Then paused, held Macet by the shoulders and at last spoke to him.

The words were repeated before he understood what was being asked. The mouth was so close to his face that he could feel the stale warmth of its breath.

'Remember the rule of this life. What is it that we must do, brother?'

Macet sobbed. 'We fall down and we get up.' The tears began to dribble down his sunken cheeks and he was babbling now. 'We fall down and we get up, we fall down and we get up, we fall down and we get up, we fall down and we get up . . .' Babbling through the foam around his lips, and with his body twitching convulsively.

Abba took him by the hand. 'Yes, we fall down and we get up.' He said it quietly, but there was still iron in his voice. 'It's a hard lesson to learn, Macet.' He helped him to his feet. 'Now say the last lines together. Then take some food from the platter we've left out. And after that some rest.'

When they were finished with the Beati, Cainnech himself almost fainted with the pain of letting his arms fall to his sides. Macet went reeling slowly back up the path, and Cainnech wanted to put a hand out to steady him, to reassure him that they were true brothers at last and henceforth always would be. But his punished body was beyond comforting the other, and his heart now shrank from fraternity in obedience and fear.

FOUR

AD 824

He was brooding over evil and sin, and the unresolved areas of grief and pain. He had been birdnesting on the ledges of the western peak, which brought to mind the first time he went looking for eggs, as a small boy. The mother, sitting on her clutch, had flown into the other lad's face in her alarm, causing him to fall out of the bush, scratched and bruised. The lad was so angry at this that he waited until the eggs were hatched, then went back with a smouldering twig and burned the nest, together with the frantic hen and all her fledglings. Enda had come upon their raw and charred corpses the next day, and wept at what his companion had done. A distance lay between them afterwards, and sometimes they fought, which had never happened before.

He could see now that this shocking event had eventually pointed him towards the monastery. He recognised, too, that he had taken the tonsure because he was also in flight from other things that horrified or deeply disturbed him in that crushingly impoverished world where his family grubbed for their sustenance from the sodden earth. Their lives did not much resemble the Hibernian ideals depicted in the old tales, where the beautiful folk always triumphed over wickedness and vice, where chiefs dealt so generously with their people that they never went short, and there was no travail that could not be salved by incantation or spell. None of the old magics had saved his grandmother—and forgive us, dear Lord, we prayed often enough for any relief—no purges or other remedies either, as she lay moaning in a pool of blood and pus, with her belly swollen hard, rocking herself tightly and crying over and over again, 'Please God let me die, please God let me die.' Nothing had restored to wholesomeness the hideously disfigured man, his nose and an ear rotted away, one arm ending in a knobbled stump, who came begging through the village one day for scraps that people would not offer decently, but warily placed on the ground before him. Dreadful as such pain and suffering was, he did not doubt that they had some place in the divine plan, which would be revealed to him in God's time, not his own. Like penance, which was also made hard to bear, often excruciatingly, they could be of redeeming worth; of that he was sure. The tormented old woman, who had herself once given him suck to quieten him when her daughter was sick and needed rest, had been brought to the stage where she wished to be with God the Son, more than anything she had conceived in her endlessly struggling life.

Enda, sitting comfortably with his back against the stone, the bag of eggs cooling by his side, watched his father in God awkwardly climbing the staircase to the saddle below, the laggard leg dragging, the shoulders knotted with pain and effort. Another act of providence, if it was true that Etgal had been lame from birth; this was their understanding, though no one on the skeilic had ever known Abba to speak of it.

There were other things that could not have been in providence, though, and Enda had early been made aware of them. Nothing in his life had hurt him more than the sight of his drunken father beating his mother almost senseless, for no reason at all that the children cowering behind the turf stack could tell; though when Enda was a little older he concluded that the man had been in despair after finding the better of his two milch cows drowned in the thin slime of the bog, and him with six young mouths to feed. But how could he have done such a thing, the boy asked himself and God, when every other night of the week he had soberly led them in praying to Christ of the tree, to snatch them from the snares of the spiteful ones.

All these years later, Enda still asked himself that question. Asked himself also whether, on those days the children came to dread, when the ominous tang of ferment drifted into the hut, and they knew that rage, whimpering fear and the sound of blows would follow soon, whether his father was then possessed by the Evil One; or whether he was simply a weak man driven by the hardness of his life to cruelty and sin. Sin was the mortal weakness that surrendered to vice, and every monk was well aware that fornication and gluttony, avarice and anger were the vices above all others to be condemned, the ones that a man in God's grace could surely choose to resist. Evil was the power that Satan disposed, a lurking presence that Christ Himself had fought in the wilderness, that had nailed Him unspeakably to the cross, a power and a presence that was unforgivable, irredeemable, because it defied even God's power to work in man on this earth. But even evil could be resisted, if not in this world, then in the life to come. Here, perhaps, it could only be endured.

Enda cupped his hands over his eyes and looked up at the lazy clouds, puffing slowly across the blue spaces of the sky. The struggle would be there, and he would be part of it. He knew that he became a monk because he had been drawn to the tranquillity of the enclosure, was in flight from the suffering and pain that dominated every other form of life he had seen. But something had happened soon after his entry into the monastic school, when for the first time he heard the story of Mhichíl and his angels fighting the dragon and his demons, and the prophecy of what was yet to come. From that moment Enda knew the rest of his life must be spent in preparing for the prophetic day, in purifying himself in the most testing way he could find in some anchor-hold, so that in heaven he would be among the army of the just. He had suddenly, and with all his heart, wished to be part of that last struggle between good and evil, in which the great whore would be destroyed, and a multitude of voices, as of many waters and of mighty thunderings, would cry Alleluia to hail the Lord God's omnipotence; and never more would a man be nailed unspeakably to a tree, or birds and their young be burned alive in their nests.

Oh, he was well aware of evil, had known it since his earliest days. But was quite unready for it when it came his way again.

The season of birdnesting had quickly passed, and it was the day in every week that came between the two fasts. Enda was down in the garden plucking weeds with one of his brothers in the hour after Terce, conscious of little but the mounting warmth of the sun on his back, the crumbs of dark soil working their way between his toes, and the grunts of a puffin from his nest hole in the bank below. He had been watching the family for days, and it had crossed his mind that no human beings he had heard of ever behaved like them. The parents were taking turns to guard the young while the other went fishing to keep them all alive.

Uneasily, he felt he was being watched himself, straightened up, but could see nothing at first. Brother Lochan was bent double with his back to him, and there was no one on the terrace above or on the ridge behind. Or flying close: birds often scrutinised their cohabitants of the skeilic with cold and calculating eyes, when perched on the buildings or riding the air just above the monks, but at this hour all except the nesters were out at sea. And then, as he swivelled upright to look beyond their cliff towards the hazy land, his stomach lurched. An unfamiliar shape disturbed the corner of his eye, was enlarged as he turned, became dreadfully recognisable when he focused on it. The high swept-up prow of a vessel was moving slowly between the smaller skeilic and them. His bowels dissolved as more of the shape edged into view beyond the slant of the eastern buttress, cautiously and silently, like a worm stealthying out of the earth; but full of menace, this. He could make out the bestial head carved on the prow, then a sail beginning to slatter in the lee of their rock, then the full terrible length of a large wooden boat. Many men standing on its deck and looking up at the clocháin, pointing them out. Metal glinting in the sun.

'Finngaill!' He got the word past his teeth, though his throat had gone dry, and his brother swung round to see what had caused the strangled sound. Looking where Enda gazed, Lochan dropped the small hoe, pressed a hand to his mouth to stifle a cry, withdrew it muttering something under his breath. So the Northmen had come to them at last.

The wonder was that it had taken them so long. They had been raiding this land for a generation now, working their way down its coasts since Enda and some of the others were children, though none of the monks had set eyes on a single Northman before. But the tale of their destruction, their plundering and the blood they shed had spread everywhere, even to remote fastnesses such as this: nowhere was inaccessible to the terrifying rumour of them. Three times in recent years the monks had strained their eyes to follow the low curving shape of their long ships distantly, as these crept close along the mainland shore, but the marauders had evidently no interest in two barren rocks so far from the normal ways of men. Monastic treasure was what they were after most of all, and such isolated skeilics did not suggest a fruitful source of that. So accustomed had the community become to being passed by, that the monks had lost their earlier habit of nervous vigilance, had subconsciously assumed it was meant that the Northmen would leave them alone. But now here they were, perhaps out of curiosity; perhaps because some Judas had betrayed the anchorites. They had found what they were looking for.

As the langskip came fully into sight, Enda's wits were returned to him. 'Quickly,' he said, and Lochan needed no urging. They scrambled out of the garden and up to the higher terrace, straight to Etgal's cell and called loudly to him. This would have been unthinkable at any other time.

'Abba, the Finngaill are here. It is time for us to go.' Enda could hear the panic in his voice. Why was he not calm, now that the great martyrdom was near? He could feel the banging of his heart against his ribs.

Movement in the darkness, then Etgal came blinking into strong light, 'Be easy, my sons,' he said. 'We shall go as soon as we are ready.' He shaded his eyes and looked to where the wide langskip was rounding the skeilic, coming to the south of them for a better view. The beast nodded horribly as the prow dipped and lifted in the swell. It was saying that their time had come. 'Ah, yes,' said Etgal, 'there they are. And they do know we are here. So we must be gathered together.'

He limped over to the leacht, where the bell had lain in its shrine since the first office of the day. He unfastened the case, put his hand through the bell's grip and swung vigorously, as though heaving a line. Before the bawling had died away, all but one of the monks were there, and he was hurrying to them through the inner enclosure door, his sandals slapping on the pavement, his robe flapping on either side.

'We shall say the Lord's Prayer. And then we shall take the precious things to the hiding place. Oremus.' Their heads were bowed as Etgal made the sign. 'In nomine Patris, et Filii, et Spiritus Sancti. Amen.' They crossed themselves thoughtfully in their turn, exactly as they did a hundred times every day of their lives. If they were to die they hoped to do so fully recognised for what they were, with everything they had been taught given its due weight and precedence even at the last. 'Pater noster, qui est in caelis, sanctificetur nomen tuum . . .' The voices were low and steady, the panic and the trembling all held fast within. '. . . et ne nos inducas in tentationem . . .' Little time left for that, thought Enda as he said it; and resolved in the same instant to do penance if he was miraculously spared '. . . sed libera nos a malo . . .' but it was not possible now '. . . Amen.'

The brethren looked to Etgal enquiringly. He sent Lochan to the oratory for the bronze cross in its radiance of sunlight, and the figure of the Saviour in a Celtic kilt. He beckoned the others to his cell and began passing things out: chrismal with the Sacrament inside; silver chalice that had been sent long ago by a king; beautiful books in wooden cases; cords so that these things could be the better carried to safety. Enda had foreseen the need for those when he discovered the hiding place. At least they hadn't been caught wholly unprepared.

'Now hurry, all of you. No, no, I need no help, I shall come behind.'

They made haste, scurrying along the enclosure and through its low door, Etgal limping rapidly in their wake. When they reached the top of the staircase, the langskip was much closer in. Some of the men on its deck were beginning to drop the sail, others were struggling with the anchor stone, more were crowded round the small boats upturned beside the mast. In a billowing downpour of hoods and gowns the monks tippled over the steps, lengthening the distance between them and the figure struggling behind. They paused at the saddle and two of them turned to go back, but their abbot waved them on as he lowered his body clumsily, laggard leg first. His voice came strongly to them across the gap. 'Go, go, while there is time.'

It will be difficult getting him even as far as the eye, thought Enda, but somehow it must be done; and God alone knows what will happen after that. The eye of the needle was his name for a vertical passage through the rock of the western peak, caused by a shattered pinnacle whose base was a vast block lodged slightly apart from the main buttress. An ascent through this passage was the only way to climb more than halfway up that precipice. He had discovered it when birdnesting a few years earlier, curiosity sending him on when prudence suggested that he would do well to turn back. To reach the eye from the saddle was hazardous enough, requiring the climber—and only one man at a time could pass that way—to face the rock and move sideways along a ledge which was never more than a foot's length in its width, and in places only half as much. It curved round the rock from the saddle to the exposed western side and, with every shuffling sidestep, the climber was more likely to fall to certain death straight into the sea far below.

The eye itself was the second great obstacle, twice a man's height and a little more than his thickness, but without much in the way of holds. Enda had been on the point of abandoning it as impossible when he remembered a childish skill he had picked up at the same time as birdnesting. The village boys would walk up the walls of two huts set close together, with their backs against one, their feet against the other, much effort and straining in between as they pushed themselves higher with each step. This had got him to the top of the eye more than once already. More vertical cliff above the ocean followed that, but there were now footholds enough to the summit, and so long as there was no wind the last part could be done without much difficulty. Just below the summit were two ledges overlooking the saddle, not easy to reach but spacious enough after what had gone before, though Enda was unsure whether there was enough standing room for all the monks. There, however, they and their precious things might be safe from attack, for even a solitary guardian should be able to stop anyone else coming through the eye of the needle: well-aimed stones would see to that. But first Enda must get his brothers past the traverse.

By the time they had struggled up to the start of it, Etgal was beginning to cross the saddle, still waving them on when they looked back. Enda decided that he must lead the first men to the top of the eye at once, then return to shepherd the other monks and be there to help his abbot afterwards. 'Remember,' he told the others as he sidled onto the sill, 'do not look down. Keep your face to the wall and watch only your step. Make sure everything is well tied into your hoods.' He had the chrismal fastened securely into his own.

No-one but Lochan had ever been up there with him before, yet only one brother became rooted halfway round the sill, fearful of moving on for several palsied moments, until Enda went to gentle him forward. 'Remember how frightening it was the first time you climbed up to the clocháin, which you now do without thinking even in a storm. It's not really so much different from that, and there's no wind round here today.' The brother nodded, opened his eyes again, sniffed loudly, relaxed his hold on the cliff face and went on. The rest shuffled round the curve with teeth bared, lips and tongues contorted in various pictures of concentration, and cheeks blown with relief when they reached a kind of sanctuary. Getting through the eye after that, as Enda had told them, was simply a matter of recalling their youth and struggling upwards at the cost of no more than aching muscles and bruised backs, with a few scratches into the bargain. Although a slip there would have killed them, too, the eye didn't feel nearly so dangerous.

After Lochan and four others had reached the top of the eye, and he had pointed the way on, Enda went back down. When he got to the start, Etgal was labouring up the slope of the saddle towards them, staggering and sliding on the loose rubble, the best of his energy spent. As Enda went rushing to help, he remembered that he hadn't seen anyone with the bell shrine. 'The bell, Abba. I think we've forgotten it.'

The crippled priest, panting hard, shook his head. 'I have it here, my son.' He took the shrine from under the folds of his cloak, a wedge of beaten silver, incised with a tracery of knots and fronds and whorls, embossed with a jewel on one side, a bronze Christ on the other. Like the chalice, it had been a gift from a pious king, who had asked only that prayers should always be said for him on the skeilic in return. Enda reached out to relieve his father of the burden, but the old head was shaken again.

'No, my son, it stays with me. It is a small price to pay if all else can be saved. Yes. They want treasure and blood. This may not be enough. I asked the Lord to guide me all the way down, and this is His will, not mine.'

'I will not leave you. No, I will not.'

'Hush. Your name is not Petrus, it is Enda, and you are under obedience to me still. Now go to your brothers and take care of them.'

A great and ugly cry from the edge of the saddle. Men at arms were coming off the lower steps, some continuing up to the monastery, others running in their direction. Hairy men with metal on their heads, metal in their hands, round shields on their backs, cloth wrapped round their legs. Terrifying in their strangeness, their savagery and their hostility. Evil.

Enda began to plead once more, but Etgal held up the hand of authority. 'Obedience, my son. Do as I say, but also pray for me. And do not let them in. Be watchful. This time will pass. God go with you all. Now hurry.'

The younger man knelt for the blessing, rose and turned away with a groan, went racing back up the hill and did not stop until he reached the sill. Lochan had also come down to help and was awaiting him, dismay twisting his face, the unspoken, unaskable question in his eyes.

'He won't come. He's going to buy them off. I tried but he wouldn't hear of it. We must do as he says. He's Abba still.'

Their last sight of him was of an upright figure standing to face the onslaught, his body hunched away from the laggard leg, as if it wished to disown the defective part. His cowl now covered the disc of his skull and its circle of white hair, but he appeared to be holding the shrine in both hands, as if he was preparing to hand over a gift. The Northmen were yelling at him as they scrambled up the slope, and some were waving their swords as though they would instantly strike him down. While part of Enda still wished to go back to share whatever happened next, another part shrank from even watching martyrdom, and was ashamed; and thankful when Lochan whispered urgently that they must go.

The shouting faded as they put the buttress between themselves and the saddle, was quite lost by the time they had surmounted the eye. A small pile of stones had been left at the top, collected by Enda when he first thought of this as sanctuary, but not added to since. Lochan went to look for more, anything loose that might be prised out of the cliff, while Enda crouched at the top of the tunnel and looked over it at the final curve of the traverse. Violence would happen here soon, which he could not avoid, and the thought sickened him. Martyrdom he would somehow, God help him, endure. It was the other thing that unmanned him more. He did not know how he could reconcile himself to offering such violence to another soul.

He did not have long to wait. In the quietness he heard someone coming above the distant rustle of the sea. Incomprehensible words were being spoken just out of sight. Then a hand with huge fingers felt its way round the rock face, followed by an arm thick with muscle as the Northman shuffled into full view, calling back to someone unseen. He was only a head higher than Enda but twice his stature, with yellow hair tangled from his helmet to his shoulders and more thickly round his jaw. His skin had been burnt by wind and light, his nostrils were wide open, and twitched as though casting about for prey. He had a broadsword at his belt; also an axe. He paused, called over his shoulder again, then turned to continue the traverse; and caught sight of Enda looking down at him. Stopped.

'Ha!' He grinned up at the anxious face, but there was no mirth in that savage mouth; and the blue eyes were icy with contempt. He shouted back again then addressed Enda with words that sounded as though he was swallowing them even as he called out. 'Berre bli staande der dú er múnk, eg skal nok fa tak i deg!' He made a vile gesture, and started to shuffle again.

Enda crossed himself, picked up a stone the size of his fist, sighted the Northman along his free arm and threw. The stone narrowly missed the man's head, struck the cliff in front of him, bounced over his shoulder and vanished. As it arrived he half-ducked, banged his head and lost his helmet, clawed with suddenly frantic fingers at the rock face, twisted his body desperately when he realised he was unbalanced . . . and fell backwards with a terrible 'Aaaah!' into the emptiness.

Enda vomited after him, almost voided completely, felt the sweat spring from every pore in his body, began to shake. And, shaking, he wiped his face on his sleeve, stooped to pick up another stone.

Hand feeling round the corner once more, questing for grip, arm and leg following, head peering suspiciously, seeing Enda: stopped.

The monk threw again, but not so closely this time. The head, the leg, the arm, the hand withdrew while voices bickered out of sight. Then silence, broken only by the comforting sounds of the distant sea and the patrolling birds.

Still shaking, Enda shifted his body, pulled another stone within reach from the pile, waited. He waited for the space of the Credo, stumbling over the most terrible part:

'. . . Passus sub Pontio Pilato, crucifixus, mortuus et sepultus; Descendit ad inferna . . .'

Afterwards, he said Pater Noster, before repeating the Kyrie so many times he could not count it. Then he began to say his prayers in the old language, spilling them excessively. 'O Lamb of God, O Lamb of God, O Lamb of God. That takest away the sins of the world . . .' And watched. And all the time was sick with anxiety for Etgal, as well as for what might happen next up there.

'Enda!' It was Lochan calling down to him. 'They are going. And they still have Abba.'

Enda shouted one of his brothers down to take his place, then climbed to the summit's lower ledge, where five others were already pressed close together against the face, like sheep sheltering from a storm. Their attention was fixed on the shields withdrawing across the saddle, and in the middle of them the limping figure, being pushed roughly along. Why had they not killed him at once? Were they going to use him to bait a trap for his children?

The Northmen and their hostage moved onto the lower steps and out of sight. Enda crawled to the end of the ledge and looked over its lip, to see if some of the raiders had remained hidden close to the peak. He could see no one, but it was best to wait awhile yet. The Northmen were said to be full of cunning as well as cruelty.

'They are destroying the clocháin, Enda.' One of his brothers pointed across the great gap separating them from the eastern peak, where figures were still moving around the monastery terraces. Some of them appeared to be hurling masonry into the sea. Well, stones could be rebuilt, so long as the builders themselves survived.

They continued to wait in silence until Lochan spoke. 'Is the net still set?' The chances were that the Northmen would not have seen it unless they had sailed close to the skeilic's northern cove before coming into view. The family would be in need of fish if the raiders had thrown the contents of their garden over the cliff as well.

As the sun began to decline, the helmeted figures could be seen descending from the terraces, empty-handed apart from their weapons. There would be a price to pay for that, no doubt, and Enda did not believe Etgal's sacrifice of himself and the bell shrine would be thought nearly enough by these devil-worshippers.

Nothing stirred throughout that night except on the western peak, where the monks kept vigil and said their prayers, and watched the stars slowly revolve across the sky to the place of all safekeeping. With the dawn, after saying the office together, they began to creep down from their refuge, Enda going well ahead to make sure they were not walking into a trap. He checked that none of the raiders was hiding in the gully that climbed to the saddle from the northern cove. The skeilic appeared to be deserted, though as they cautiously looked to the south, they could see the langskip still riding at anchor in the same place as before. But it had all its boats on deck and, unless this was part of a ruse to capture the rest of the monks, that meant the raiders were preparing to leave before long. Three of the brothers went up to the monastery and found no one there; but all the cells had been damaged and some were tumbled down, while the leacht and the oratory had been disgustingly defiled.

Enda and the others were turning onto the lower steps when they saw him. There was a huge block of stone set apart from the skeilic, much too far away for a man to leap onto, yet almost too close to sail a curach in between the two. Because it slanted massively towards its parent, it was known to the monks as the Leaning Rock. The

Northmen had placed their captive on this. Etgal was standing crosfigel in prayer when his brothers found him, and he seemed to be
still unhurt. He was half-turned away from them, more towards the
langskip, as if he was making intercession on behalf of his prosecu-
tors. And though Enda knew this was very possible, he also realised
that the abbot at that moment was oblivious of the vessel. For he was
simply facing the sun, which was now well above the mainland hills.
Mutely, Enda prayed with him, while his eyes ran watchfully over
all the approaches from the landing place. 'Flame of a splendid sun,'
he implored, 'apostle of virginal Eriu, may Patrick, with many thousands, be a shelter to our wretchedness.'

The Northmen saw the monks before Etgal did. Most of them had
been sleeping under small tents on deck, but the watchman noticed a
movement on the cliffside above. Stooping, he picked something up
and held it mockingly at the monks, who saw it was their bell. When
he swung it, the high clang carried over the water, at which men were
awakened and clambered to their feet. Some began to jeer and wave
their arms as if to invite the Christians on board. No one attempted to
launch one of the boats, though.

Etgal did not move through this clamour. He stood there on the
brooding black rock with his arms apart, as if willing the Maker to
draw him up into the everlasting peace; but a little while afterwards
he turned and faced his brethren. He gave them his blessing across
the space between them, and they bowed obediently in response.
Then he returned to the sun and continued to pray. Throughout that
day he followed the sun with his prayer, his body moving with its
course until at last it dipped into the western sea. Always facing the
sun, indifferent now to the skeilic or the ship or to anyone they contained, separating himself carefully from everything on earth. He
was standing when darkness fell, and on a moonless night the monks
could not see him until the new day began. He was still at crosfigel,
awaiting the sun.

Enda wondered what the Northmen were waiting for, why they were doing nothing but sleep and eat and wait. Occasionally they brandished the bell and its shrine at the monks, laughed among themselves, sometimes played a counter game on the deck, waved at Etgal tauntingly. But did nothing else, while the monks kept vigil and watched Abba prepare himself.

The Northmen had underestimated how long it takes to starve a man to death when he has spent a lifetime training himself to go without food. Nor did they know—how could they have known?—that a man of faith can will himself to be with his god, if his faith and his will are powerful enough, more powerful than his instinct and desire to survive. In that scant body, made to endure by being broken from birth, Etgal's nature struggled with its contradictions for eight full days, always in the light of the sun. On the ninth day, cloud covered the heavens so thickly that only a cheerless glow was visited upon the earth. When the next day arrived, the abbot was no longer standing, but was sprawled across the rock, still in the position of the cross.' A gull was stabbing at his neck hungrily, but he never stirred. In the middle of that day, the Northmen put down one of their boats and two of them climbed the black stone to inspect their captive. They picked up his slackened body, held it for a moment so that the monks could see; then dropped Etgal in a misshapen heap, his limbs lying unnaturally. 'Father forgive them . . .' The words formed themselves in Enda's memory; but he could not get them near his heart or past his lips. As soon as the boat was stowed aboard the langskip again, her anchor was raised and she set sail for the land. The Northmen never once glanced back at the monks, who could only ponder the subtle cruelty that had kept the Finngaill there so long.

How could these heathen possibly have known the nature, the magnitude, of Christian guilt? Yet why else had they stayed, while doing nothing more than wait for their abandoned victim to die? Enda at last understood that he would never be free of the stain, as long as he lived. Etgal had not even received the viaticum to help him on the way to where he would be, but there was much more to repent on top of that. Enda now knew that fornication, gluttony, avarice and anger were not the vices above all others which condemned themselves. Betrayal was that. He would forever be haunted by the most awful of all the questions crowding in on him; one he would never be able to resolve. In which of his treacheries had he committed the gravest sin?

AD 950

As the greatest of their fasts drew to a close and the most sacred day of all approached, the monks gathered their sluggish energies together for the transformations of the holiest week. Throughout that Lent they had denied themselves fish and, taking but little food from the garden when they ate at all, each of them was aware that his body was weakening just as his determination grew to be properly prepared for the great sacrifice. Tasks were accomplished as before, but gradually took longer to perform. Words were imperfectly chanted at the offices and so penances were endured. Men found their attention wandering in the smallest exchange with their brethren, requiring all their tenacity to hold fast to the impending truth. Even sleep became more elusive, as the longer vigils and increased penances and reduced nourishment began to take their toll of the body, making it restless, without ever numbing an imagination that seemed to stimulate itself through all their exhaustion. Some of the monks therefore ignored their beds on fine nights, exchanging the sweet herbal scent of dried plants for the sharper cleanness coming off the sea. They took themselves to the small terraces they and their forebears had constructed all over the skeilic, where a man could be apart and undisturbed by his brethren, to dwell upon the Maker in his own fashion and in his own time.

Ciarán's hermitage was a turfy bank with a little wall he had built with his own hands, overlooking the sunset cove where the seals had made their home. To reach this from the saddle, it was necessary to clamber past pinnacles and buttresses of rock, outcrops of the towering western peak. These formed a natural enclosure behind anyone who sought solitude halfway between the sky and the sea. Keen ears were necessary below this bastion to catch the ringing of the bell that told the office hours, and Ciarán was occasionally required to do penance because the wind had carried the sound away from him on a dull and blustering day. He was never late when the days were bright, because then he could judge the time himself by the shadows cast upon the cove.

There he would sit motionless for hours with his legs crossed and his huge hands upon his knees, and allow the wonder to soak into his senses until there was room for no more and he was bursting to shout of it in ecstasy and praise. His first anamchara, many years ago, long before he came to the skeilic, had taught him to sit upright at these times in order to be alert to every sign that the Lord might make; and, after half a lifetime, the posture had become so habitual that Ciarán's body was able to relax only as an act of obedience. He was himself now the soul-friend of a young monk, who sometimes wondered whether part of Ciarán had not already left them for the kingdom to come. He would be explaining something, shaping visions with those long-fingered hands, and his voice would trail away, and he would look past the young man into a distance beyond what was visible; and into his eyes would come a look of great contentment that no one else on the skeilic quite possessed. Others carried the hardness of certainty in their glance; but there was something more persuasive in Ciarán's gaze. It sheltered almost shyly beneath the eyebrows, which came together in a great tuft above the nose, and were so thick and dark that they were much more prominent than the greying coronet of hair higher up his head. Someone had once remarked, when Ciarán was growing up, that it was strange to find the softness of a doe in the body of a bull.

He spent hours on his terrace meditating on the skeilic itself: on its substance, on the variety of its elements, on the significance of its shape. He had chosen this monastery above all other island fastnesses because, as a youth and finding the way to his place of exile, he had heard that this skeilic rose from the sea in the semblance of two hands close together in prayer. At once he knew that this was where he must eventually be: there was a rightness and a trueness that went to his heart and was affirmed there without the slightest hesitation or doubt. Never would he forget, as long as he lived, and lived again, his first sight of the skeilic as he came to it across the undulating autumn sea. For the dream he had dreamed for so long was at last a reality, and there was the rock rising just as had been foretold, with its fingertips gathered together and pointed to heaven. Over the years since that day, when out fishing in the curach, he had seen every one of its different outlines, and none of them had failed to present some form of that devoted image. No one could ever have doubted that this was meant to be a holy place of incessant prayer.

The skeilic was also miraculous in its variety. The Leaning Rock was square-set and almost entirely black, and all the ledges of the skeilic itself which were low enough to be regularly washed by the waves or sluiced with spray were likewise dark; but above them rose buttresses and galleries and ridges and spikes and sharp edges of stone that foretold to Ciarán a divine purpose which would move the world from darkness to light. Many of them changed colour as they responded to the sun, when a drenched morning was followed by brightness later in the day. Sombre and ominous when lashed with rain, they would glow with warmth as the first beams pierced the cloud, and if a full blazing glory came afterwards, they could be seen in a silver majesty of their own which itself radiated light. And on these unimaginably ancient witnesses of time were more recent marks that measured the passage of the seasons. Even the steps laid by the monks—was that yesterday?—were now dappled with silver and gold smudges and smears that had spread over their grey or dark blue surfaces with the years; and few were the higher rock faces of the skeilic without a growth of green beard, or without clumps of pink flowers thrusting from their seams and crevices.

When Ciarán sat on his terrace he felt that he was surrounded, protected by and in communion with living stone, which would call to him and perhaps tell him the secrets of creation—for these rocks had been formed here on that day—if only he would learn how to hear. Every time he came, he prayed that he might hear the mysteries of the Beginning, know the full upheaval, the cataclysm that Genesis barely outlined, as plainly as he could already see the semblances of life in these snouts of stone, in sharp profiles and fallen gargoyles of men, in an entire bestiary outlined in rock; a baying hound, a crawling animal with a tusk and a squat figure riding upon its back, a rearing head on a long neck that for some unaccountable reason—for he had seen nothing like it before in his life, nor heard it described—made him uneasy in its representation of sinister power. There were sharp flakes of silvery rock that overlapped each other like the scales of a fish. It was as if the creatures which took refuge in the Ark had been petrified here in everlasting testimony to the inundation of the Flood.

Below and around was that self-same sea, which could sometimes terrify with its irresistible power, and at others could soothe the soul as much as the softest morning in the lush pastures amidst which Ciarán had been born. He had known it at its most violent, smashing and sucking hungrily at the skeilic as if it would devour the very rocks, flinging itself so high up the cliff that it showered the clochán with its spray. But the year so far had been unusually calm and, as this week began, the loudest sound was the echoing boom and swash of the waves as they thumped into the nooks and crannies at the waterline.

On the ledges of the skeilic's western tip, the seals now lazed for most of each day, heavy with fat and full of fish, crooning to each other as their instinct turned to new life. They had long since reached an understanding with the monks, who left them alone, in return for which the seals ignored the net so long as it was not set across their sunset cove. There was more than enough fish to satisfy them all.

It was the birds that thrilled Ciarán more than anything else, and not only because he longed to imitate their flight, to drift over the surface of the waters with leisurely beating wings like the gainéad, which would presently climb much higher than the skeilic and then, folding itself into a pointed shape, would dive onto a deep swimming fish so perfectly that it scarcely left a ripple on the surface. Approaching middle age, the monk still longed to imitate this; or he would wish to soar halfway to heaven with the staidhséar and the other gulls, gliding ever upwards on by-blows of air, then sideslipping down in an immaculate swerve to come skimming past the clocháin with the speed of the long-haired star. But it was not this enviable miracle of flight that so enthralled Ciarán, nor yet the other miracle that enabled these creatures to fill the sky with myriads of wings beating, fluttering, adjusting, twitching, bodies hurtling, turning, swooping, dropping, without ever, without once flying into each other or into anything else they had not wilfully decided upon. Even the puifín, which made Ciarán shake with laughter when it tumbled over itself in its haste to rush from its burrow into the air, even the puifín never faltered once it had risen from the skeilic's platforms and grassy banks, and went paddling off into the wind in its daily quest for sustenance.

It was the sheer joy of the bird-flight that captivated him most. On Palm Sunday that week, after they had made Christ's body on the leacht at the mass, Ciarán went to his hermitage for the time that was left before the later offices. He was wondering whether the seals, too, were meditating in spite of their apparent drowsiness, when a schisty cough caused him to look into the sky above the cove. The black shape of cág cosdearg was circling overhead, his ragged wing ends pressing the air down with confident strokes, flexing as the boy Ciarán's hands had done wishfully long ago when he first became entranced by flight. Cág's mate then appeared from behind the western peak, drew alongside, almost nuzzled him, and in perfect harmony the two birds swam through the sky effortlessly and in sheer happiness. There could be no other explanation, thought Ciarán, as he watched them soar together, deftly change places, simply disport themselves. They were not seeking food or nesting materials, they were not on passage to some other place. They had taken to the air, it seemed to the monk, for no other reason than the enjoyment of each other's company in the exhilaration of the powers they had been granted in the Beginning according to the Word. It was as though they were trying to show him something about their being, its divine origins and its heavenly purposes. They were using the cove as an amphitheatre now, surely conscious of their audience, flying high above the encircling walls, then dropping in somersaults almost to the level of the waves before rising again with wings full splayed and wedge tails steering them away from all harm. Twice they climbed higher than before and, as they reached the summit of their ascent, they turned over on their backs and plummeted as if they were one until, with a huge upward swing of delight, they swept past Ciarán's terrace so closely that he could almost believe those curved red beaks were grinning with pride.

He would not have heard a hundred bells summoning him to prayer at that moment, when he sensed that he was on the brink of some profoundly new understanding of creation, which would lead him to know more clearly than before the Maker Himself. And from that he could hope to understand what was intended in mankind's relationship with these creatures, these rocks, these plants, this water, this air, this unity of all things on earth that offered glimpses of the divine. Of something grimmer, too, Ciarán was well aware, for the creatures perhaps most of all, but in ways that were also in the experience of man. The immaculately headlong dive of the gainéad into the sea left many of them blind as they grew old, so that they could no longer find food and perished miserably. Only a few days before, he had surprised a falcon which had just struck from the sky a young gull, and which scolded the monk angrily when he interrupted the beginning of its meal. It returned later to pick up the gouged and spreadeagled carcass and carry it to a nesting place, leaving only a twist of pink gut and a scattering of curly white down upon the path. Ciarán knew man to be the greatest hunter of all, as merciless as any bird of prey, who even on this skeilic, this dedicated and holy place, took fishes without remorse, and eggs that were meant to become young, and spared the full-grown birds and the seals chiefly because he lacked fire to make them palatable. Ciarán had never been easy at this. If their God was indeed the God of the sun, the moon, all stars and all life, as Patrick had assured everyone who followed him, why were so many of God's creatures made the victims of man? Had Christ Jesus come into the world to save sinners from these iniquities, too?

If there was an answer it passed from Ciarán's understanding once more, with the passing of the two birds out of his sight. But he carried the question and pondered it daily, through all the terrible and tremendous happenings of that Holy Week.

As the Passion unfolded, the monks redoubled their devotions and intensified their fast. Since Lent began they had endured with Christ the wilderness, subsisting on no more than a single stalk of greens three days in every week. But after the deceptive triumph of Palm Sunday, nothing but water passed their lips until the miracle of Easter had occurred. At the same time, their life became even more focused round the leacht, their normal labours suspended until the great fast was done: all save one. As Holy Week began, the net was set again across the northern cove, and after two tides it was retrieved with three mature bradán and five smaller fishes in its mesh. Gutted and split and hung amidst leaves to protect them from the birds, these were strung up between the clocháin to be prepared by the wind for the feast that would follow the Easter mass.

Otherwise, Abba and his disciples concentrated themselves wholly on Christ and His followers, day after day. They chanted their psalms and read the Gospels as always at the offices, but in their overwhelming awareness that these were the very days on which the sacred drama had been enacted almost a thousand years before, they were possessed, as if without the power of self-control, by the deepest of all their desires, which was to do homage and to submit themselves. The prostrations they made at every act of worship throughout the year were reticent compared with the times they now went to their knees until their heads touched the ground, or lowered themselves completely and lay crosfigel upon the slabs of the terrace surrounding the leacht, in patterns of absolute humility. And so the offices lengthened until there was little space between each, but in these small intervals the monks contemplated what had just been done, and what was yet to be, in that week long past but always now.

When the chief priests and the scribes sought how they might take Him by cunning, Ciarán wondered which of his brethren could possibly have been or be Iscariot. When it was his turn for Abba to kneel and wash his feet on the night of the Last Supper, he had an urge that he suppressed with some difficulty, to reach out and embrace that bent and devoted head, to take the tender hands in his own and raise up the humble old priest. At the trial he felt sick with apprehension, and when Pilate had finished interrogating Him, sentenced Him and turned Him over to the mob, he began to tremble with the old fear of agony and humiliation. At the Crucifixion, he felt pain in his hands and his feet, excruciating pain as if he had touched fire or had himself been pierced, was aware of the tunic beneath his gown sticking to the wound in his side but, when he looked at his body afterwards, was surprised to see only sweat pouring down, without a trace of blood. The most terrifying moment of all came when He thought the Father

had deserted Him, and Ciarán did not dare to think what would become of him if that truly happened at his own end. He struggled
through the emptiness and desolation of the following day in a grief-
stricken daze, and once thought his legs might be about to give way
under him.

The week was calm and dull from Palm Sunday on, but occasional
shafts from the sun had begun to slant through the cloud, so that the
shadows on the ocean beneath were broken by pools, splashes and
wriggles of light, the waters glistening where the light fell, quietly undulating and gloomy where no direct light was able to reach. Easter
Eve was black with only a thin sickle moon occasionally visible, when
Ciarán left the monastery and picked his way slowly through the pinnacles and buttresses that lay between the saddle and his hermitage.

He tried to still the turmoil by breathing carefully, as he had been
taught to do by the same guide who urged him always to sit upright.
He emptied himself of all thought but the thought to pacify his body
and his emotions, and allowed the night air to drift in gently until he
could accept no more, holding it for a moment or two as if it were the
fullness of a tide, before letting it quietly ebb away from him. At first,
the fullness caused the inside of his head to flash with sudden bursts
of light, but after a while these died away and his breathing was so exactly measured in its ebb and flow that his body scarcely stirred at all.
He remained like that, suspended in senselessness, for so long that he
never noticed an increase in the stars and the drawing back of cloud.
He was listening to God, not talking to Him, waiting for a sign that
would confirm all the truths in which he believed: the oneness of
everything in creation, the redemption of mankind in Christ, the life
everlasting and the adorable transformation of the bread into the
Body at the mass.

He did not stir until long afterwards, when he realised that the sky
was no longer impenetrably black everywhere, but on the verge of
greyness overhead. Something—was it Him?—made him look up
to discover this change, his first movement after becoming utterly

still. He realised, too, that the grief had passed from him in the night, that instead there was tranquillity and, yes, there was also eagerness. He could feel the eagerness beginning to course through his veins, beginning to warm the cramp from his limbs. Stiffly, he gathered his body together and climbed up to the saddle so that he could watch the dawning of Easter over their anchor-hold. When he had clambered through the enclosure of pinnacles and buttresses he stopped and leaned against a rock, where there was a clear view through the gap between the skeilic's two peaks. If the day was going to be cloudless, he would soon be able to see the mainland coast. Already the eastern sky was beginning to pale above a dark blur that could only have been the distant hills. The pallor changed to whiteness, which swiftly acquired warmth. The faintest hint of yellow became primrose, then buttercup, then something close to marigold, and as this happened light flooded across the sky so that what had been invisible was made visible at last. Ciarán could see the ridge above the clocháin now, with some of his brethren standing or on their knees, also waiting for the sun to rise.

And then it appeared, in the blink of an eye, such an astounding event in such a tiny mote of time. A particle of redness, no more than a hairsbreadth of fire, was there on the horizon, starting to come above the roundness of hills. Almost at once its fieriness curved; and in that instant Ciarán blinked again, then rubbed his eyes, to be sure that they were not fogged with sleep. For the fieriness at this dawn was like no other he had seen before, at Easter or at any other time. This sun was not rising into the sky as it normally did, ascending steadily without varying its course, promise of power so absolute, so terrible, so incredible that soon it could no longer be looked upon. As it moved upwards, the sun this day was also moving inside itself, throbbing as though something within was bursting with abundance of life. There, just above the horizon, on this day of days when Christ rose from the dead, the sun was pulsating rhythmically like the heart; or as if to the beat of a drum and the note of a pipe.

In that instant Ciarán recognised the sign for what it must, could only be. The sun was at the heart of all things, and it was dancing with joy for the risen Son!

He did not doubt it from the moment the vision formed. It could be nothing else. The Son was risen and with the Father. He was risen indeed. And the sun was a sign to all mankind that this was truly so.

Ciarán let out a great wild cry, flung his hands above his head and brought them together in a loud smack. He could see his brothers on the ridge moving like children in their excitement, pointing, waving to him, beckoning, clapping in return. One of them (was it young Brusc, to whom he was anamchara?) tried to somersault in his happiness, but toppled over and slithered down the ridge. Not a thing to attempt on an empty stomach, perhaps. Ciarán himself needed a steadying hand on the rock, after a last look at the whitening heat in the east, which already hurt his eyes and made his head swim. Well, the fast would be over by the time that sun was at its height. Now it had resumed its normal path, moving upwards much more quickly than any star, almost as swiftly as the waxing moon. The puckered sea was becoming as if molten under the heat, a golden column seeming to rise from its depths to the white heart of the sun, a great pool of light spreading everywhere across the waters.

But before the feast there was the mass to celebrate the greatest miracle. The sun was triumphantly above their heads when the monks gathered at the leacht in the lenient warmth of that midday. They had confessed their sins soon after the dawn, each man telling his soul-friend of wandering thoughts, of infidelities to their way of life, of failures in charity, all his weaknesses; and on this day alone they had been forgiven without reservation or penance because Christ had just taken all sins upon Himself and had died for sinners everywhere. They were cleansed and without stain; only bound to Him the more in perpetual gratitude.

They therefore celebrated the great mystery with reverence as well as with joy. No other day in the year ever radiated as much happiness as the Easter morn; and at no other mass did they ever feel quite the same unassailable certainty that the Presence was there, at the leacht, in the chalice, beside the altar cross of bronze, enfolding each one of them. On this Easter especially, they were aware as never before of mysteries, of miracles, of the goodness that would ultimately vanquish all evil, of the unity in all creation, of the shining Godhead who empowered it. As the abbot intoned the first of the liturgical sentences, Ciarán felt himself treading closely in the footsteps of the first apostles, stronger in his faith than he had ever been. The promise was there in the Credo, and as he came to the heart of it with his brethren he knew these were the holiest words that had ever been said:

'. . . Tertia die resurrexit a mortuis; Ascendit ad caelos; Sedit ad dexteram Dei Patris omnipotentis . . .'

The chalice was uncovered after that and the old priest uttered the syllables of consecration. Ciarán began to tremble so much that his teeth were chattering when it came his turn to receive. Abba placed the Host in his mouth, then pressed a careful hand upon Ciarán's head. The trembling passed. The stale crumb on his tongue, laden with so many hopes, so much pain and suffering, so many verities, became the sweetest, the most vital thing in all his life; the one necessity. He knew that he could die for this; and that he would probably die if it was ever withheld from him.

Much later, after the monks had come as close to gluttony as they would ever know, after they had feasted themselves upon the fruits of their garden, and the wind-dried salmon, and sucked from their shells the molluscs they had picked at that day's low water, and after they had said the remaining offices, Ciarán returned to his hermitage for the last of the light before the darkness returned. But, he reminded himself, the darkness and light to Him were both alike, for He was the Light that transcended the darkest day the world had ever known. And in this Light anything was possible, including the endurance of agony.

It was even possible to see a harmony in the coexistence of tremendous opposites and contradictions; in the breathtaking dive of the gainéad which resulted in the death of a fish, in the unyielding resistance of the skeilic to the dreadful power of the storm, in the dedication of men and women to the Creator by renouncing their own ability to make life. In this harmony there was gloom everywhere, but it was not endless darkness, only the passing shadow which intensifies the light; there were setbacks but no defeats; the desolations were less than everlasting tragedy. Since the Son came among us from the Maker, there was always and eternally illumination and hope. Not being afraid of death or other endings: that was the almighty secret. Believing that something came after them: that was the truth. Accepting whatever transformation was meant: in that was the promise of eternal life. Christ had come into the world not only to save sinners, but to reconcile opposites.

Ciarán watched the sun going towards the sea, a few high clouds radiating the brilliance of its light in crimson fishbones across the sky. A breeze had stirred the waters a little earlier, but now it dropped, as if the world was holding its breath in awe of this second transformation of the day. In the morning there had been resurrection, and now there was something else; but whatever this was, it was not an eternal death, for tomorrow the resurrection would come again. As the ball of shuddering fire slid down into the sea on the very edge of the world, there was a last flash of radiance in the heavens above, a final manifestation of the great miracle; and then it was done. But the Light would return; of that he was sure.

The breeze stirred again, and Ciarán drew his cloak around him as the shadows closed in. Before he picked himself up to go back to the monastery, he wondered what it would be like at the instant of his own passing, at the time when he ascended into heaven. He did not doubt for a moment that he would be with Christ one day, though all else was unclear to him. He imagined himself soaring into an emptiness much like a bird, and looking down at the skeilic's upturned fingertips, unable to see his brothers but quite sure that they were wishing him well and praying him on. The skeilic's own form of prayer comforted him, too. What came after that great pause in his life he could not be sure of yet, not in all its parts, not the complete and awful immensity of it. And, for now, he was content to contemplate it as a mystery.

SIX

AD 1044

After the Culdee was gone, they began to understand something that might have occurred to them when he arrived. He was preceded by strange omens, which had been unknown within living memory. No rain fell that year between Epiphany and the morning after Ascension Day, more than four full months of what was normally the wettest season, a drought so prolonged that, by the end of it, one of the monastery's cisterns was dry and the other was very nearly so. The monks had at first welcomed relief from the customary rain storms, which also enabled them to watch the low winter sun crouching naked throughout its course for day after day. But they had become perplexed after Easter, when clouds were still rare and without moisture, and when they realised that their garden crops would be stunted if no rain came soon, that their water supply could very well run out before Pentecost.

That was also the spring of the black cloud when, more than halfway through Lent, the sun was obliterated in the fullness of its brilliance: not suddenly, but gradually, by a darkness that crept over it and made the shining noon so eery and ominous that they suspected this might be a time in which demons would breed. Not one of the brothers had ever heard of a cloud as powerful as that; nor had they known one that was invisible before and after it obscured the sun, and announced its coming only by casting unaccountable shadows across the sea. There had been a short but violent storm just before the stillness that accompanied the extinction of the sunlight, and another brief tempest after the sun became visible again. Had this happened on Good Friday, or at any time in Holy Week, the monks would have been in no doubt about its significance. As it was, they were simply troubled by something inexplicable that mysteriously threatened the natural order of the world. Not one of them would forget it, to his dying day.

And then the Culdee arrived and he, too, was destined to remain always in their memory. He came to them after the drought had been well broken, on a midsummer's afternoon of great warmth, across the twinkling sea in a butty little boat, which he managed alone. Only one of the brothers had seen a curach like it before, somewhere in the far North, on the coast near his childhood home. No one noticed the Culdee's craft until it was passing the little skeilic, partly because it was so small, but also because the nervous habit of intense vigilance, learned at great cost in one after another of the Northmen's raids, had become unnecessary. The invaders had eventually conquered the mainland and made new settlements there, but had sometimes married the native Celts and accepted Christianity; even one of their kings, it was rumoured, had sought baptism. The raids on the skeilic and other monasteries had ceased. The Northmen who held themselves aloof from the Gaels, however, had never been seen as anything but a curse upon the land; and only a few years earlier they had been defeated by the ard rí Bóruma, in a great battle which lamentably cost him his life.

So the monks had become accustomed to regarding any craft upon the nearby sea, whether coming in their direction or not, as peaceful and friendly in its intent. And apart from the boat that every few months brought them their necessities, other curachs did occasionally visit the skeilic now, when the sea allowed in the lighted half of the year. Always these had a solitary passenger, some sinner who sought to cleanse himself by doing penance in this holy but notoriously un-comfortable place, and then to be absolved by the monastery's father in God. The penitent was always warmly received, questioned about his sin and, when the abbot had determined how long the penance would last, the boatmen were told when to return and take him back again. Once or twice every year this happened nowadays, the visitor living as a member of the community, attending the offices with the monks, sharing their labours and their food; and in addition per-forming the prescribed penance for his particular offence in whatever time remained to him each day. Some, it was said, were transformed by the experience, living holy, contrite and otherwise exemplary lives on the mainland afterwards till the day they died. None had ever been known to return.

Even though it was unprecedented for one man in a boat to reach the skeilic, the monks assumed this must be another penitent when the curach was close enough for them to see that it contained a soli-tary occupant, pulling vigorously on the oars, as he must have been doing for several hours. Three of them accompanied Abba down to the southern landing place and patiently waited there for the visitor, but he never appeared. A shout from above presently told them why, one of their brothers indicating that the curach was about to land in the alternative northern cove. The four monks therefore climbed back up to the saddle and down its other side, but before they reached the cove, the figure below them had already stepped ashore. He was lean-ing over the curach with his back to them, pushing the little boat out to sea again.

Frowning now, the abbot quickened his step as he led his brothers down the path, and as the curach began to bobble away from the skeilic on the flooding tide. The stranger had turned to meet them with clasped hands, and bowed deeply as Abba approached. He was a tall man, clad much like them, with a tonsure that had as much silver in it as black. His head sat on him with jaw held high, not the carriage of a naturally humble man, and the mouth looked as though it had been bred from generations of disdain. Dark eyes were downcast as this monk acknowledged the abbot's rank; but otherwise they looked out severely over the bridge of an eagle's nose. Something curiously between a small birthmark and a bruise discoloured the skin of the forehead just below the hair.

'I am Aedh and priested, and I am come from Terryglas,' he said. The voice was quiet but strong, with a trace of impediment that might easily tantalise.

'Ah, Terryglas,' Abba replied. 'Then you are welcome. But why have you . . .' He gestured towards the curach, which was now well away from their rock.

'I have no further need of it. It is not intended that I return. I do have permission, Abba. And, I hope, yours.' He stooped to open a small satchel and withdrew a scroll.

The abbot knew the reputation of Terryglas, the monastery beside the lake, not so very far from Clonfert and Clonmacnois, a stronghold of the Culdees, who had arisen some short time before the Northmen began to raid, dedicating themselves to a renewal of the religious life, which they deemed to have become lax in some of its observances. The abbot had once, as a young man, thought it possible he might himself take vows at Terryglas or even Tamlachta, but the Lord had purposed him otherwise, and he had never regretted this. He doubted whether either of those houses lived more closely in the way of the Desert Fathers than the monks of this skeilic had unswervingly done for many generations now. Moreover, the Culdees, he had later realised, were perhaps too proud for his taste; for he was more watchful of pride in himself than he was of the carnal sins, because he knew it to present the greater hazard to his soul. They thought of themselves as the Companions of God, but they allowed themselves to be known by other vanities.

The scroll was from his fellow abbot, and made it clear that Aedh was here, as he said, with his superior's express permission. It commended him as one of the most pious and disciplined of the Sons of Light—there it was again, the pride of them—who had served his monastery faithfully and well these twenty years, growing daily in the spiritual gifts. But, wishing to set himself an even sterner test of his worthiness to be with Christ—well, that was certainly a rare compliment from such a source—and after much prayer and meditation, he had come to the conclusion that the remaining years of his life on earth must be invested in the skeilic, that his hopes of the life to come must henceforth he professed from there. In which the abbot of Terryglas unreservedly concurred. He added that he was sure Aedh would be an ornament to his new community.

Abba stifled a snort as he came to the end, and hoped that his smile would be thought merely amiable. It was self-will, of course, that had caused Aedh to cast off his curach before he was welcomed into the community. A very determined will it was, too, which had also observed the state of the tide and its currents, and then calculated which side of the skeilic to land on, so that the boat would be carried away at once.

They were to discover that this was invariably Aedh's way, though no one would have gone so far as to accuse him of false humility. But Abba at once decided to make himself the Culdee's anamchara. He foresaw that it would be a challenging relationship for anyone else, perhaps undesirably so. If Aedh was indeed highly disciplined as well as pious—and this, after all, was a famous distinction of all Culdees—he was not going to split disputatious hairs with his superior. But his first confession, which the abbot heard on the day of his arrival, made no mention, showed no awareness, of pride. With expressions of deep remorse, however, Aedh sought penance for the sins of thoughtlessness and negligence in casting off the curach, when it might have been valuable to his new community. He was, he told Abba, afraid that he might be sent away again if he had retained the boat.

'No one coming in good faith has ever been sent away from here. And no one ever will be unless we all find that . . .' The abbot needed a phrase that would not hurt this supplicant.

'There are difficulties?'

'Yes, difficulties. A key to this life, as you know, is learning to love those you do not naturally like. Some find that very hard. Some find they cannot love in that very special and generous way, as they are meant to in a monastery.' Abba remembered just such a case, when he was a novice at Clonfert, which had resulted in so much disquiet for everyone else that both monks had been asked to pursue their vocations separately as hermits, at some distance from the monastery. This was rare, but it happened. He grimaced apologetically. 'And then,' stretching his hands, 'we should all have to decide what to do.'

He wondered why he was telling Aedh this. He had never spoken of such a thing before to any of his brothers.

A shadow of distress clouded the patrician face. 'Abba, I would never allow myself to be a source of difficulties. As long as I live, I want only harmony with my brothers under your authority. With that help, I seek only Christ.'

'Of course you do, my son. Therefore there will be no difficulties. And now you must reflect that another key to this life is submitting yourself to God's will without reservation, putting yourself at hazard to Him. For your negligence in that, you will go to the oratory now, and on your knees you will recite the psalms before you eat or take rest. Now go in peace with my blessing.'

It was not until the middle of the night, a little while before Lauds, that Aedh completed the Psalter, with a final exultation of praise which left him close to ecstasy on the oratory floor. In the short time that remained before the office, the abbot, who was a light sleeper, was disturbed by the swish and thud of a body being disciplined. He did not venture outside his cell to see who the flagellant was: he did not need to, because the one scourge belonging to the community, for use only under his advice, lay in a corner of his room. The Culdee had evidently brought his own.

When the brethren were gathered at the leacht, the abbot was aware of another sound, unfamiliar but not unlike the impact of knotted cords on a man's back, which occurred at the office every time the prostrations were made. Unable in the darkness to discern what caused it, he awaited the dawn and their next act of worship together. And at Prime he could see that at each prostration the Culdee lowered his forehead until it struck the ground, not clumsily but as a deliberate act. So: that was how he had acquired the curious mark on his brow. Well, mortification of the flesh was well known to produce great sanctity, though the abbot was uneasy whenever he remembered that blessed St Ita had allowed—had seemed to encourage even—a beetle to eat away at her side without complaint.

Later, when it was all over, he would bitterly regret that at the outset he had not dealt more firmly with Aedh in the matter of flagellation. But he had been consciously tentative, reluctant to seem overbearing so soon after welcoming the newcomer, especially when he was a fellow priest.

'Is it wise,' he had asked at the Culdee's next confession, 'to scourge yourself quite so frequently? In imitation of our Lord's suffering? It is very easy to lose sight of His face when we form habits which become ends in themselves. When we stop asking ourselves what He would have us do.'

'I do not use the whip in imitation of Christ's Passion, Abba. I have . . .' He cleared his throat, and for the first time since arriving seemed ill at ease.

'Appetites?'

A small and hesitant nod, with lowly eyes. The abbot drew breath in the silence, thought very carefully what to say next. There had never been a monk untroubled by his appetites, and pain was only one way of reducing them. As a young man, he had always found that prayer face downwards on the cold stone floor of his cell was an effective form of self-control.

'Be very careful, Aedh. Nothing should become an end in itself. We need to be disciplined in all we do—even in the use of the discipline.' He smiled at his own pleasantry, and allowed the moment to pass.

But he should have been firmer than that, should have exercised his full authority on the Culdee. Had he done so, he reproached himself long afterwards, things might in the end have turned out differently. The flagellation, he reflected, had been the first station on this via dolorosa. He sometimes wondered why his brethren had chosen such an indecisive creature as their father.

It was months before the next stage was reached. And in that time the monks found their new brother perfectly agreeable. If there seemed to be little warmth in him, an aloofness even, that set him further apart from everyone else than some might have wished, there was certainly no other cause for complaint. The Culdee was assiduous in all his duties, unsparing of his energies, whether cultivating the garden, shifting and repairing stone, or cleaning the terraces; and he was ever ready to help the three old men, who nowadays took longer than the others to complete their tasks. He was courteous to everyone, the first to step aside when he and a brother approached each other along a narrow path, the one who insisted, with a graceful turn of the wrist and inclination of the head, that others should go before him to take food from the plate. Invariably he was the first to reach the leacht after the office bell rang, he never faltered by so much as a syllable in the chant, and they soon stopped listening for, found that they no longer noticed even, the punctual bump of his forehead on the ground. Everything Aedh did increased his reputation among them for great piety, notable even among men whose whole existence was an aspiration to holiness and nothing else. Except to his confessor, however, he gave nothing of himself away but his energy. Not one of his brothers was ever to learn anything from his face that wasn't written there the day he arrived: no passing sorrow, no joy, no exasperation, no surprise, no doubt, no perplexedness. Only self-control and serene certainty.

One day it struck the abbot that, for all the superiority of his position, he was becoming intimidated by Aedh. This was the one thing that scrupulous man did not reveal to his own anamchara, the oldest monk on the skeilic, who would probably have counselled some form of humiliation for the Culdee: something that deprived him of sleep, a penance that invariably sapped all forms of stubborn self-will. The abbot had no wish to exact punishment when his own pride was so manifestly at stake; besides which, he felt sure that Aedh would emerge from the experience with his reputation not only intact but probably enhanced. This would not threaten the abbot's position, to which he had been elected until he day he died, but he feared that it might subtly undermine his spiritual authority. He recognised the weakness of this fear, and meditated long upon the nature of self-deception and other forms of dishonesty.

But he connived at the beginning of the separation, and at everything that came afterwards. Aedh had taken to spending more nights away from his cell than in it, but it was only the number of these absences that was unusual. Individuals had always been in the habit of periodically keeping vigil apart from their brethren, as the Spirit directed them, but they had remained firmly rooted in community. Now and then someone had felt the need for an extended period of prayer and fasting and reflection in solitude, for as long as a month sometimes, and permission was never refused for one man at a time to create his own temporary hermitage. It was understood that sooner rather than later, the solitary would return to the coenobitic life of the terraces, refreshed and eager for all the old rhythms of community.

Aedh spent his nights of absence at the far side of the saddle, where a strange collapse of slabs had formed a small cave. When he sought a blessing for this, the abbot questioned him about his need.

'What impedes you in the shared life with your brothers?'

'Nothing impedes me, Abba. But there is something beyond it which I may need to reach.'

'In all humility, I hope.'

A lowering of the head; assent, something more than mere gesture, perhaps.

'What is it that you seek in solitude?'

'I do not know, Abba. I know only that there is something be-yond . . . a vision, a possibility. Something more. Something I sense in solitude.'

'And you will . . .?'

'I do not know that either.'

Such certainty cloaked in unknowing. And the abbot himself was sure that Aedh would not return to the clocháin once he had left them behind. There was something else as well as self-will driving this man, and he did not think his own authority sufficient to stop it, even if he wanted to. But in truth, he would welcome the end of nightly flagellation within earshot of the terraces, which was beginning to dis-turb the tranquillity of some brethren, causing a small distraction that lay somewhere between impatience and envy. One of the dangers he had feared was drifting down on them.

So he gave Aedh the permission he sought. And, again, the Culdee threw himself wholeheartedly into such life as he still shared with the rest of the community. It was at his suggestion that they built the new chapel, and no one laboured to finish it more devotedly than Aedh. It had been the tradition to say all offices and the mass at the open-air leacht, except when the weather was so torrentially wet that not only would it have soaked them to the skin in a few moments, but would also have drowned the sound of their canticles. On such days they worshipped in the shelter of the oratories, a second one having been built long before their time because of this very need. But it was an unsatisfactory solution, to be divided thus; and even with the two buildings, thirteen monks were so cramped that genuflection was al-most impossible, full prostration out of the question.

It was after Sext on a day of uninterrupted downpour, that Aedh went to the abbot's cell and stood bowed in homage at the door until his superior appeared and invited him in. Always he showed the most proper deference. Continued to do so as he wondered aloud whether it might not be propitious to build a church that would accommodate them all with ease in bad weather, leaving the oratories for private devotions, as was the intent when the first of them was made. 'A church would also,' he finished, 'be of service if pilgrims were to be drawn here, Abba.'

The abbot looked at him sharply. 'This a place of pilgrimage?' he asked. 'What makes you expect that?'

'It is possible that this may happen one day, Abba. Pilgrims are on the road throughout Christendom now. Not only to the Holy Land, to Peter's shrine, or to St James in Galicia. Pilgrims now take sustenance from many other places that have been touched with holiness. One day they may come here, too. They will bless us for building them a church.'

So they began to build, in a space between the abbot's own cell, the larger oratory and the cashel wall separating the upper terrace from the garden. It took them almost a year of quarrying and dressing and fetching the stone, then placing it, block upon block, with two doorways and a low ridged roof, which was daringly wider and longer than the two already constructed for the oratories. And no monk's hands were torn more than Aedh's with working the stone, no member of the community whose heart was more obviously given to finishing the chapel. It was Aedh who chiselled the piscina and placed it outside the west door, where it would be naturally replenished with the holiest water in the world, raining down. It was Aedh's proposal that they should dedicate their chapel to the archangel Mhichíl, who had, he reminded them, become the patron of high and holy places in many lands. And on the saint's festival, they held their first mass there, with a collect of consecration which Aedh had composed and which he recited before Abba made the miracle with the bread. As this day came between the Lents of Moses and Elias, they feasted themselves afterwards in honour of their new patron saint.

It was on the first anniversary of the dedication that Aedh announced his purpose to make the separation complete. 'I think it is time, Abba, for me to withdraw from our communal life,' was all he said. No great deference now, only absolute and controlled certainty. Perhaps he knew that he needed no permission to leave. The abbot had been prepared for this; was relieved that the break was now to be made.

'And where will you go?'

'Up there, Abba. I do believe it has been waiting for me.' He was pointing to the summit of the western peak. The all but inaccessible place.

The abbot had expected him to say that he would leave the skeilic by the next boat that called, to seek some uninhabited other rock for his hermitage. But this was an almost laughable conceit; except that the Culdee had never once been in danger of provoking mirth. The western peak had been used as a refuge at various times during the years when the skeilic was threatened with attack from the Northmen. Monks had also now and then climbed to its summit out of curiosity, but none had done so without great difficulty on the calmest of days, and none had ever shown the slightest inclination to spend an unnecessary night up there: the hazards were all too obvious, from the steepness of the rock, which rose from the sea like a wall, from the extreme narrowness of the few ledges, from the power of the wind perhaps most of all. The abbot's single experience of the peak, shortly after his arrival on the skeilic, had unnerved and humbled him, crouching like an animal at the top of the needle's eye when a sudden and violent rain storm almost swept him from his fingerholds. It was strange how one could look forward to death, yet remain terrified of the last steps to it: and it occurred to him that the Culdee might conceivably be indifferent to even the most frightful possibilities. No, his proposal was by no means laughable. It was staggering in its audacity.

'How can you possibly dwell up there in solitude?'

'By building, Abba. By making a little garden and cutting a small sink. But only with your blessing and with the assistance of my brothers. And afterward, only in the knowledge that the prayers of everyone down here are upholding me.' He bent his head to acknowledge his dependency.

Oh, he was clever, as well as utterly determined and sure of himself.

The building of the hermitage engaged all of them, including the old men, who could no longer climb rock faces or even, in this case, attempt the difficult traverse that led to the needle's eye. What they could do was to dress stones, and make and repair the things monks had always needed for their construction work on the skeilic, the leather bags and ropes with which the stones could be hauled to wherever they were needed. It was only the younger brothers who on all but the foulest days of that year went up the peak in between the offices, manhandling the stone and placing it where Aedh directed them. For the first time that any of them could remember, his face was no longer impassive, but alive with an excitement that he did not attempt to conceal. His eyes, one of them remembered afterwards, now had a gleam that intensified with every stage of the construction, shining with a passion that might have moved mountains; or could equally have destroyed anything that stood in his path.

He had worked it all out long before Abba gave his permission. He had been climbing the peak regularly, without anyone thinking it significant even on the rare occasions when he was noticed, in the months of his interregnum on the saddle. Before the first building stone was hauled through the needle's eye, Aedh had made a start on his hermitage. With a chisel that he had brought with him from Terryglas, but never shared with anyone, he had chipped grooves down the face of the summit, and then hollowed out of the rock, just below, two small basins for his water supply. These were on the widest ledge he could find, on which he now began to construct his oratory and his leacht. On a well-turfed lower ledge close by he would grow his food. On the outermost ledge of all he would create a little shelter, and from here he would each night face the setting sun; and know, one day, that his own time was almost here. This was the most precarious place on the entire peak, apart from the tiny tabletop of the summit itself. It was when watching Aedh on his knees while he carefully made a wall there along the very edge of the abyss, that one of his brethren was dismayed to find himself wondering whether the Culdee might not be possessed by the same spirit that had tempted Christ. All the ledges needed low walling to prevent things sliding off them into space; but this one was by far the most dangerous for a hermit to choose as his principal anchor-hold.

The monk later regretted that he never confessed his anxiety to the abbot or anyone else.

When all their patience, their industry and their fraternal concern had created this defiant aspiration to perfect holiness, and the hermitage at last awaited its occupant, the brethren held an open-air mass to fortify the Culdee in his new life. It was a fine day which reflected the optimism of spring, and on this feast of Beltaine their forefathers would have driven their cattle through circles of fire and danced round these while facing the sun. As they celebrated at the leacht, the monks were not unmindful of their deepest heritage, or the profound instincts of their ancestors, but on this morning their thoughts were focused on the tremendous pillar of rock which rose behind them, pointing the way to the eternal hope and the ultimate truth.

They gave him all their goodwill that day, as he remade his monastic views before them and sealed these with the Suscipe. His voice was as distinct and steady as it had ever been when he thrice chanted the words: 'Sustain me, O Lord, according to thy promise that I may live; and let me not be disappointed in my hope.' The 'Amen' from his brothers was as confidently loud as any the terraces had ever heard. They knew, each one of them, that after this morning they would never again be close enough to him to touch. He was going away from them; yet he would always be within reach, visible or invisible up there on the peak, a sign of complete devotion, a promise that nothing was impossible, and an encouragement to all who would live with integrity, however strange its form might seem to some.

Aedh did not look back as he walked away from the monastery with his satchel on his shoulder, possessed as he had never been by anything before. At last he was free; free to be what he had always known he was meant to be. A man set apart, a Sign of Signs. After so many years with a dream, now began the reality of becoming a hermit whom no one could possibly imitate. He was to live like John of Lycopolis, at the end of a path which was barricaded to stop the curious from disturbing him. But the curious and even the indifferent would hear of him in time, because his brothers knew of his devotion and would henceforth be witnesses to every sanctified day he lived up there, beyond their reach but not outside their regard. He would become the chief of the Gaels in piety, and legends would grow around him as they had been handed down from the life of Patermuthius, for whom one day the very sun stood still. His own piety would be no less real than that of the early Fathers, whose lives he was now free to imitate perfectly. It pleased him to think that he might, like some of them, enjoy the companionship of other creatures on his crag, that some bird might respond to his every call in time (he had also seen beetles up there and, once, the trace of a mouse). Otherwise, he was well content to dwell alone with his vision of the life to come, and his careful preparations for the call to be with Him.

He was, almost completely, independent of every other human being now. His oratory was visible from the upper terrace of the monastery, and it was understood that if he hung his satchel from its eastern end, one of his brothers would that day leave some precious bread in a bag below the needle's eye, so that he would always be able to celebrate the Eucharist. But the Culdee made sure he never encountered these helpers, and he needed them for nothing else. He had transferred some of the plants from the community's garden to his own, where they flourished in the lighter soil once the first rainfall had watered them. He missed the fish, but fed rather better on the vegetables than he had been accustomed to in community, where excess in all its forms was frowned upon, and brothers privately competed with each other in the severity of their fasts. He was intensely happy with the small labour involved in the gardening, and in keeping his stonework in good repair.

He quickly acquired the habit, after his first experience of a storm, of climbing round his boundaries to check that none of the stones had been dislodged, none of the plants uprooted or bared. Finding a thin slab left over from the building of the hermitage, he began to chisel a ring cross on one of its sides, a simple, unembellished outline, carved in all humility; and when repair or tilling was necessary, he fell upon it, too, as a blessing, muttering prayers and murmuring hymns as he worked: 'Oh bright Sun, that illuminates heaven with much of holiness. Oh King, that rulest angels. Oh Lord of men . . . Heaven's might, Sun's brightness, Moon's whiteness, Fire's glory, Lightning's swiftness, Wind's wildness, Ocean's depth, Earth's solidity, Rock's immobility. Blessed be thou in the name of the Lord . . .'

Now and then, he giggled uncontrollably.

The focus of his life changed over the next few months, from the formal liturgies to spontaneous prayer and other forms of contemplation, though he celebrated mass every day now, instead of every week. He could still hear the distant clang of the monastery bell for the offices, but he did not always respond to its summons: sometimes he would remain sitting cross-legged in the sunshine and continue his dialogue with Mhichíl, or Christ, or the Maker Himself. But if he ignored an office, he later signalled his omission by making a devotion that he had learned at Terryglas, saying the Pater Noster to the east with hands raised to heaven, then to each of the other quarters, then with head bowed low, and finally with face held upwards to the sky.

Sometimes, exhausted after hours of prayer on his knees or standing crosfigel, he would sit relaxed against the mountain wall, and look out to sea until the visions began. Angelic figures that were both tender and voluptuous beckoned him to ecstasies that he had never known but did not want to forget, and now that he was alone he surrendered himself at last: he used the scourge to punish and not to subdue. He more than once thought to imitate Origen afterwards; just once took up the razor and ran his thumb along its blade. Guilt in what God had given him.

As the old year turned into the new and winter stormed in from the nothingness beyond, Aedh spent more time in the oratory or in his shelter on the edge of the abyss. It was far too dangerous to stand upright on his outer terrace in the full fury of an ocean storm, which could be so violent that when the sea smashed into the skeilic from the west, it then climbed up the cliff and reached the hermitage with its spray. He once spent a day huddled in his hut, defying the waters to take him down to their depths. He knew they could not, because Almighty God had need of him at the right hand; and he laughed with exultation at every crash that jarred his dwelling to its foundations. Then the weather eased as spring came again, and soon all the skeilic below him was smothered in white as the campion bloomed, while pink flowers sprouted in all the cracks. On the calmest days now he would climb to the very top of the peak and sit for hours overlooking everything in sight; the great gap of the saddle separating him from the eastern ridge, the domes of the clocháin snuggled along the ridge's slant, the little skeilic skirted with white water beyond and, in the hazy distance, the green breastings of the land, curve after curve of them, with dark circles of furze where no sheep or cattle had grazed. He felt, up there, as though he need only reach a little higher to touch with his fingertip the outstretched hand of God. And he knew that what he saw now was as nothing to the revelation awaiting him, when he was himself risen and able to survey all the kingdoms of the world in one glance.

More visions were sent to him. It was Holy Week again, and he had eaten nothing for so long that he was a mere ruckle of bones with skin opaquely covering them where flesh used to be. He was finding it difficult to rise from his prostrations, and even from his knees. He did not dare crawl to the outer terrace now, and was passing the time till the Easter feast beside or within his oratory. Just before sunset one evening he lay slumped against its doorway, when figures appeared over the retaining wall, inhuman creatures which had the appearance of headless men, but with eyes and mouths set into their chests. They were followed by people with ears as big as the sail of a curach, which they wrapped about their thin bodies to keep themselves warm. Men and dogs' heads came after, and a black apparition which climbed over the wall on four human legs and began to crawl towards him on all fours. He closed his eyes to see Christ and made the sign of the cross, and when he looked again the apparition had gone; but a man with one huge leg was sitting on the wall instead, sheltering from the sun's light beneath his foot and ten toes. Everywhere Aedh turned now, he was surrounded by creatures that he had not even imagined in the underworld: a man's head with three rows of teeth, set upon the body of a lion, but with a scorpion's tail; a horse with a goat's jaws and long straight horns sticking from its skull; two ants the size of mastiffs, great talons on their feet, with which they were digging for gold; a bull in a field of its own endless dung, which reeked of smoke and crackled with flame. Some of these creatures were merely strange, others were also terrifying; yet even the figures that were frightening as well as out of this world were banished when Aedh made the sign of the cross. They were to recur, though, even after he had broken his fast and recovered his strength.

The abbot had thought to invite him to their Easter, and himself took bread up to the eye that week, with some half-dried fish in case the Culdee preferred to feast alone. He had hoped that Aedh might come down in greeting at his approach, but seeing nothing of him, he let it be and returned to his monastery. The brethren caught sight of their hermit often enough, a distant compelling figure against the sky, and were conscious of themselves being watched from up there. The monk who had worried about Aedh felt this so acutely that he sometimes turned swiftly to see the Culdee holding them in his eye, quite motionless. In the early days of his withdrawal, some of the brethren had waved a hand in fraternity whenever they saw him, but he never returned their salute and eventually they stopped offering it.

He very rarely noticed them, in fact. He was generally studying the birds: their motion through the air; the way their wings controlled their flight; the flexing of their legs as they pushed themselves into space from a ledge. But he was not unmindful of his brothers below; he prayed for them every day as Mhichíl's locum, guarding them against all evil and purging them of all sin. And soon he would be empowered to do the same for the whole world. He became utterly sure of this after the worst summer storm he had experienced, which battered at the skeilic so terribly that he thought its purpose was to pluck him from the hermitage and take him to eternity that very night; until he realised that he was being allowed a little more time to make his preparations complete. He had never known a storm which arrived so unexpectedly and which blew itself out after a single day. The night after it raged was almost silent, and a half moon hung in the sparkling sky. When he awoke, a great blanket of fog covered everything except his peak. In every direction was nothing but a rolling white cloud, and the sun blazing across the stainless air; no monastery, no little skeilic, no mainland; nothing at all but himself, purified at last, and the sun above the clouds. That morning, he knew the bidding time was very close. He had indeed been singled out.

He was very methodical in his preparations during the rest of the day, as he had been all his life; in his coming to the skeilic, in establishing his hermitage. While the sun was burning the fog away, he used every drop of his water to cleanse each part of himself, so that he should be worthily immaculate. He celebrated mass and gave himself the viaticum. Before he climbed to the capstone of the peak, he raised the thin slab with the incised cross and stood it upright, so that all who came afterwards would recognise how this place had been touched by true holiness.

It was a quiet day still but now one of total clarity, beautiful beyond the saying so, a promise of what was to come. He knelt for a while high above his old world and repeated his Creed for the last time on earth.

'. . . Inde venturus est iudicare vivos et mortuos . . .'

He was ready for the judgement now, and for all else that would follow it. He looked across at the monastery, and could see his brothers turning away from their office at the leacht. They had never really known him down there, had never quite accepted him. He smiled indulgently. The mystery of myself, he thought, is the mystery of God. He looked up at the sun, and no longer needed to shield his eyes from its light. Its light contained Christ's face, which was smiling at him, bidding him come without further delay, to sit with Mhichíl at the right hand. He raised his arms and held them there while he studied the length and slender breadth of them; arms of silver, much stronger than they looked, perfectly made to take him to God.

'Look!' said one of the monks, pointing towards the peak.

The naked figure was standing crosfigel upon the summit, its head turned to the sun. Then the arms flexed and the legs bent a little, and the body was in the air. Some trick of the wind, a powerful draught rising from the sea perhaps, carried it upwards at once on gracefully outstretched hands, but only for the time it took to draw a breath. Then the arms flailed frantically, not with the confidence of a bird, but with the desperation of a drowning man. And the body plunged out of their sight in an awful and uncontrollable whirling of limbs.

Exaltation was suddenly transformed into terrible awareness, appalling truth. Slowly enough for the truth to horrify him, the grim rocks and the seething of the sea rose up to meet his fall. His head was split open, from tonsure to chin, his brains slattering out like the ordure of a gull. And in that instant he saw before him at last the yawning chasm of his wilfulness and pride.

SEVEN

AD 1222

The fatal storm merely made up Abbot Eoghan's mind that the time had come for them to take the next step: the idea had been forming for a year or two before it happened. The world had changed since his saintly predecessor first brought monks to the skeilic, and every one of the changes had at least distantly affected the life of the little community. The grey men from Gaul had long since spread themselves across the land like the Northmen before them, their iron-meshed bodies impervious to the weapons of the Hibernian tribes, whose own unprotected ranks were cut down like sickled corn by these new invaders. The best the clansmen could usually hope for was to take refuge in a tower, whose door was too high above the ground for an attacker to reach, and to stay there until the danger had passed; but such relief could only be temporary. In all five kingdoms, the chiefs had been squabbling again, with the invaders at the old game of playing one off against the other; and the Church had also been taking sides.

Rome, in any case, had been in an expansive mood for years before the grey men arrived, extending the power that she had first exerted in the age just after Columcille. The monastic order no longer retained its traditional primacy here, having been obliged to submit its abbatial authority to bishops pronouncing ex cathedra. Mael Maedóc O Morgair had been Rome's chief instrument of what it pleased Innocent II to call reform, in return for which he had secured high papal office to go with the mitre he already wore in the North. And it was true that some abuses had crept into the monasteries, all needing correction; not only the selling of pardons and other forms of simony, but cases of men holding the abbatial title to two or even more houses, which demonstrated that they were not true monks at all. (Even viler things had been perpetrated against the Church's property, the monastery of Inisfallen having been plundered of all its worldly wealth by Mael Dúin, son of Domnall Ua Donnchada, and only God's mercy had prevented him from killing people in this heavenly place or stealing its sacred books.)

Mael Maedóc had also brought back from one of his Roman excursions some of the white monks from Bernard of Clairvaux's foundation, and settled them in at Mellifont, the first of many aliens who would be transplanted here now that a start had been made, with all that meant for the exercise of monastic authority. It had become clear that such control would pass from pure and freeborn Celts in their native land, to a number of hierarchs in distant countries where only mongrels bred: they would probably see Benedict's black monks in Munster before long. Eoghan did not doubt the faith or the sincerity of these men, only their founder's own commitment to the example of the Fathers, judging by the accounts that had come his way. Benedict had said he was setting up a school in God's service, but it sounded to Eoghan as if he had created something close to a temple of sloth and gluttony instead. The notion of unbroken sleep for most of the hours between dusk and dawn, of three substantial meals and a measure of wine every day, passed the understanding of anyone who had lived thirty-eight disciplined years as a monk on this skeilic, his devotion to Christ burnished more brightly with every day that he hungered and thirsted in his fidelity.

Single-minded holiness in the footsteps of Antony had withered even among proud Celts, however, and with it the burning zeal of individuals to ready themselves for God wherever a desert could be found. Fionán could probably have filled a fleet of curachs with aspirants when he set off into the great emptiness. But it had been many years since Eoghan's community equalled the full number of the apostles, and none of the brethren now were truly young men any more. Three of them were very old; so old that they had long since carved the small crosses that would eventually be placed above their graves. Yet although the lifeblood of the monastery flowed sluggishly these days, the attraction of the skeilic to others had only quickened in recent times. It was said that nowhere else in the whole of Eriu did pilgrims come in such quantity to seek spiritual sustenance from anchorites. There had been occasions, even, when they outnumbered the incumbents for a week or more, straining the ability of the monks to be both faithful to their eremitical vows and hospitable to their guests. Things had now reached the stage that in calm weather, when it was possible for a boat to make a landing, brothers who should have been totally absorbed in their devotions, kept a distracted eye open for some curach heading their way with one or more penitents aboard. There were moments when Eoghan believed he was beginning to experience the tension his predecessors had lived with permanently when the Northmen were prowling here.

Clearly there was a need, a deep hunger in the increasingly disordered world across the water there, for the certainties that were signified in the stability of their life. But what if the stability became threatened by the hunger and the need of others who ultimately depended on being sustained by men living this particular call to holiness: what then? Should the hungry be turned away to obtain what strength they could by simply being aware of the holiness at a distance, without being allowed close enough to feel sanctified by themselves touching the hem of its garment? Or should the stability itself in some way be modified, so that the monks might become more accessible, even if this meant that they would therefore become a different kind of monk? Eoghan ached with something deeper than sadness when he thought of their backs being turned on all that had been witnessed here in the past six hundred years, all that had been accomplished in Christ by scores and scores of Fionán's children, generation upon generation of them; this especially difficult and utterly unstinting gift of themselves back to the Maker, that only very few of His people had ever been able to make. To renounce that past would be a form of treachery, to be sure; but to imagine renunciation where there was none would be a folly as shameful, if the intention were truly to bring others more certainly to the kingdom of heaven. And, after all, none of Fionán's children had ever been wholly independent on their skeilic, had ever been able to cut themselves off completely from others in Christ. Not one generation of monks, who had needed a boat every few months to bring them necessities. Not even the proud and pious Aedh, who had needed the help of his brothers to make his hermitage and then to supply him with the bread of Life.

Eoghan had turned these matters over many times, had filled his meditations with them for hour after hour over a number of years now. He had come to the conclusion that perhaps their vocation must take a different way in the path that would one day lead each of them to Christ. The way forward now might be to make their home on the mainland within sight of the skeilic, there to live a modified form of the life which would be more accessible to penitents. The skeilic would become a place of retreat for those monks who needed to withdraw from time to time for a period of more rigorous self-discipline. It could otherwise be a place where pilgrims were allowed to practise their religious exercises, directed by one or more of the brethren. There must, of course, be strict control of all these comings and goings. The example of Aedh had been a painful one, which still stung even though he had passed into legend. No abbot since then had licensed anyone else to run such risks on the western peak. There would be no more totally unsupervised zealots courting spiritual disaster on the skeilic.

Eoghan, as cautious as any of his recent predecessors, and having come to these conclusions, then spent months re-examining all the possible alternatives simply because he shrank from being the one who ended such a long and humbling tradition in this holy place. He had even summoned a chapter of his brethren in order to share the burden of his responsibility, but it left him no nearer setting a date for the upheaval. The old men, the ones who would not see threescore years and ten again, were even more reluctant than he was to turn from their island home, and not only because it had become comfortable to them. Others, too, wanted to retain the purity of their vocation and could not see how this might be if they shifted elsewhere, when they would be indistinguishable from any other kind of monk. Only two of the brothers could see some virtue in moving, and they without enthusiasm. Yet not one of them demurred, not even the most inflexible; they also bowed their heads in obedience to Eoghan's design. Submission had been everything in their lives: to the will of God, to the example of the early Fathers, to the direction of Abba. Generation after generation of them had known no other way. It was why they were so different from other men, who rebelled, relished their anger, insisted in indulging self. Therefore in accordance with their father's wishes, they decided to hold the future even more in their prayers, and to await the sign that would surely come, as it had come to blessed Fionán.

So it was the storm which settled things. It had come upon them from the emptiness of the west, where most of the worst weather began, for reasons they could not possibly divine, any more than they could fathom why the sea was salty whereas inland waters were not, why the tide ebbed and flowed, whence the winds arose, where thunder and lightning came from, how and why the globe of the earth was held up in the middle of the air, and why lightning was always accompanied by thunder, though thunder often occurred alone: all these happenings being in the nature of God, by Whom eventually all things would be revealed. The storm was yet another of His wonders, its beginning perfectly familiar to them, its climax both unprecedented and terrible. Its purpose was all too manifest.

Winter had blown itself out and the first flowers had started to colour the green banks when, one day, the wind arose again and the swell of the sea began to increase. Where there had been small troughs among the waves, there were now widening and deepening spaces between massive slopes of water, which were enlarged until they were as spacious valleys sunk low among the hills. As the fierceness of the wind grew, it first made great waves along the upland ridges of these swells, which broke and crashed down the rolling slopes in a rush of foam; then its fury increased until it tore the waves from the ridges before they could break themselves, and flung them through the air to be smashed against the slope rising ahead. Such seas as these they had sometimes seen before; the assault made on their anchor-hold was utterly outside their experience. Before the storm spent its worst, they had seen the Leaning Rock disappear time after time under mountainous seas, had been transfixed as a column of water rose up the western peak and almost over its top before collapsing upon itself, were terrified when something hugely ominous flew through the air over their heads, which turned out to be an entire bank of turf torn from its bedrock by the wind. For two days the monks remained apart from each other in their cells, telling the offices and saying their other prayers, feeling the walls shaken to their foundations and making themselves ready lest they should fall, hearing the almighty violence around them in trepidation and in awe. When at last they ventured outside, they discovered that some of the enclosure walls had been blown down, with part of the chapel roof collapsed. Some of the crosses in the graveyard had been scattered away.

They had also lost three of their brethren. Two monks at the outset of the storm had gone down to the cove to haul in the fishing net, and had never returned. When the others became alarmed by their absence, two more went off to look for them, one running faster than his companion towards the staircase leading to the saddle. As he slithered from the shelter of a buttress halfway down the steps, he was snatched by the wind as if he were a doll and flung from the rock far out into the sea before his brother's horrified eyes. Those three missing men were the youngest of the monks, the ones whose vitality could be spared least of all if the community was to continue as before. Their bodies were never returned from the depths, though Eoghan and his remaining brethren searched each cove and rock pool and prominence every day, until they finally left the skeilic in submission to the will of God.

They left on an easy morning just after the Feast of the Transfiguration. From the moment on Easter Sunday when Eoghan declared that they themselves must rise again elsewhere, each of them had spent many hours wandering round the skeilic, lingering over its familiarities, touching the roughness of its stone, stroking its tufts of green beard, bending over its new flowers and rustling them gently with the backs of their hands. They would stand or sit for long moments gazing out on the unpredictable yet changeless sea, or looking up at the peak, whose inscrutable stone mutely waited for time to pass, or watching the fortunate seals and birds, who would still be here long after they had gone; wanting to treasure it all in memory. Every one of them tried to remember how that pillar of rock, that view of the clocháin, that curve of waters surging into the cove, that distant sweep of coast had seemed when first he came here and took it all in. The slap of the waves on rock, the echo of a bird's cry against the cliffs, the whisper of the wind up the gullies, suddenly seemed to be the saddest sounds in all the world. But after a while, they wanted to be done with the letting go, wanted to get it over, wanted to be away to their new life. It occurred to Eoghan that it might be much like this if he became very advanced in years, like Simon and Aedóc and Olcan, their three old men: something that he would feel as well as, but apart from, the burning desire. He was going to be the first of the abbots whose bones would lie somewhere else. A stab of pain when he thought of that.

They repaired the broken parts of their monastery, they restored the fallen crosses above the graves, they swept clean their cells and the terraces, they made bundles of their precious things and other necessities, they said their final offices at the leacht. With everything ready for their departure, they celebrated mass for the last time. Eoghan wore the full liturgical dress of his rank, gifted by a king five generations earlier, which the abbot normally wore only for the high feasts; amice and alb, girdle and maniple, stole and chasuble, a pattern of faded green and red, or material that was no longer white but the colour of old bone, with gold thread laced through much of the cloth.

Slowly they made their way down the steepest path to the east landing, with a gentle breeze coming from the west that would push them to the land. Not a word was spoken as they climbed down, and each monk paused now and then to look back, pretending to have an eye out for Eoghan, who came last of all: but it was not really so. The curach was tethered, awaiting them, the bundles already stowed, and they helped each other across the tackle and the thwarts until everyone was in his place. They crossed themselves without a word as the rope was untied; then the oars were put into the water and they pulled themselves away from the cliff. Facing it as they bent over the spars, they studied its features closely for the last time: the ledge of the landing, which was always treacherous with weed; the steps leading away from it, which also needed care, chiselled out of vertical rock; the boom of the sea inside a cave, and the way it seemed to be draining out of all the skeilic in little downpours and gushing spouts; the white splashes, hundreds of them, beneath all the nesting sites; the deep unvarying green of the turf banks and the many greys of bare rock; the enclosure wall and the peeping domes behind, visible once they had pulled away from the great overhang; the terraces, see, which every man knew as well as he knew the inside of his cell, and the dog-leg of the main staircase as it came down the south side from the saddle; the shrieking, quarrelling, circling, soaring, diving of the birds, which were staying on. The everlasting things. But they were leaving it all now, the distance between them had already begun. They could see it slipping away; six hundred years of history, their history, their faith's history and the history of their people; they were turning their backs on that. That was the pity and the great heaviness of it all.

When they were well clear of the lee, and past the smaller crag too, they shipped oars and paused before preparing to sail on. It was so many years since any of them had been so far from the skeilic, that they observed it now almost as a novelty, this rock that had long been their home. The ones with good eyesight could still see the clocháin, stuck like limpets just under the eastern ridge, on a field of green; the old men could only make out the slant of the skeilic, the shape of its prayer, with the thumb of the western peak sticking out.

'The Lord will take them up into Himself, won't he, Abba?' asked one of the monks. He had been looking at the sea rather than their rock, and Eoghan had guessed what was on his mind. 'They will be able to find their way from out here?'

'Of course they will, Colmán. There's no end to the mercy of God, and however hard it may be to see it, we must believe that this too was merciful. He wanted them to be with Him, as well as to point the way for us.'

'Young Padraig had an expectation of this, I think,' said another. 'Don't you remember, Abba, him saying that he imagined it would be as easy to resurrect from here as from the earth, with the soul leaping upwards like a fish?'

Eoghan nodded. He did remember it well. The three of them had been standing on the terrace on a perfect evening about the time of Lugnasadh and the young monk had been entranced by the light gleaming off the waves. He had said that he could not imagine a finer resting place than there, body as well as soul purified by the immaculate and cleansing sea.

'He was a great young sun, was Padraig,' said Olcan, his rheumy old eyes brimming with memory. 'So were they all.'

Eoghan nodded again, absently this time. He had been trying to count the days since that first night of the storm. 'I do believe,' he said at last, 'that they may be there by now.'

He looked round at his brethren, and saw them smiling for the first time that day. The curach had suddenly filled with tremendous happiness.

They stepped the little mast and hauled up the sail, and at last turned themselves to face the land. The day was so clear that earlier, from the monastery, the coast had seemed close; they had even been able to make out shadows falling across the inlets between headlands of its cliffs. But now, from the level of the sea, it appeared to have receded from them, or shrunk to a line of humps which ran unbroken as far as the eye could see, from north to south. Having almost no knowledge of what they were going towards, Eoghan simply chose a notch in the land that, at the moment they raised sail, happened to be directly under the climbing sun, and began to steer for that. Much later, when the sun was at its height, the curach was running into a bay, with low cliffs on either side, and hills sweeping upwards just beyond: green hills, but their soil was perhaps very thin, for there were many stumps and jags of stone sticking out of the turf. Ahead was a line of low rocks, with a passage through the middle of them, and a wide stretch of water just beyond. Eoghan steered for the gap and, once the curach was through, he had the sail and mast brought down. It must now be the way it had been in Fionán's time, in the will of the Maker to choose where they should be.

They drifted straight ahead for a little while, until the snout of the boat began its turn to the nearer shore. Slowly they approached a strand of sand and small rocks that sloped gently upwards from the tide. Beyond this was a flatness of green before the land began to roll, and beyond that was the steeper shelter of the hills. The place of their resurrection must be over there, then. Eoghan could see it clearly now.

As the boat rocked quietly towards this shore, the abbot broke their silence at last. 'Let us say what we believe, what will always go on. Oremus.'

Eight heads bowed, eight voices repeated their creed, the words of faith that had been kept on the skeilic these six hundred years and more, that had altered and transformed it for all eternity. They knew that well; the skeilic would never again be as it was before Fionán led men of God to worship there. It was impregnated with devotion, with prayer, with another form of holiness that matched its own. It was a token, a pointer, that humankind would never be able, or want, to ignore.

'. . . Credo in Spiritum sanctum, Sanctam Ecclesiam catholicam, Sanctorum communionem, Remissionem peccatorum, Carnis resurrectionem, vitam aeternam. Amen.'

As the last word was said, the boat touched the shingle spit. The monks began to help each other ashore, passing their tackle up to the grassy level above the strand. Later, Eoghan and one of his brothers returned to the curach and sailed it back to the gap between the low rocks. There they cast it away from them on the outgoing tide, setting it free to find its own destiny and to be another sign. Until it met its end, it would be a sorrow upon the sea.

PART TWO

The Evidence

*The number above each section of Part Two
is a page reference to Part One.*

3
Fionán

In attempting to reconstruct the history of the skeilic, one of the greatest difficulties lies in identifying the monastery's founder, Fionán. At first sight, a number of candidates seem to suggest themselves, for there were several men of that name (or its variants) during the early years of Celtic Christianity. The most famous of these was the Finnian who is generally seen not only as the begetter of the regulated monastic life in Ireland, but as the instigator of the great tradition of monastic scholarship there. Living in the age immediately after Patrick's, he studied for many years under David and Gildas in Wales before returning home to found his own community in 520 at Clonard, by the headwaters of the River Boyne in Meath, close to Slighe Mor, 'the great road' running from the east coast to the kingdom of Connacht in the west. Clonard became the exemplum of the Irish monastic school, its most celebrated pupils including Columcille (Columba), who later founded his own monastery on the island of Iona, and Brendan the Navigator, whose name became attached to some legendary voyages. Because so many of his students were eventually regarded as men of exceptional holiness (as he was himself), Finnian of Clonard has been remembered above all as 'the teacher of saints'.

Another mighty figure was Finnian of Magh Bile (or Moville), who is said to have been a novice under Ninian in Galloway, before having to leave in haste when a Pictish girl fell in love with him. There then followed a period when rumour had him studying in Rome, before he returned to his native Ulster and there founded the community whose name is associated with his own, near Newtownards. This Finnian is credited with having brought to Ireland from Rome a biblical text in Latin; and he, too, was certainly a scholar-monk like his namesake of Clonard. According to one of the traditional accounts, he is also supposed to have had a quarrel with Columcille which resulted in the latter's going into exile, and was thus indirectly the genesis of Iona's foundation. Yet neither of these Finnians can seriously be entertained as the founder of the community on the offshore skeilic, for a number of reasons, one of which is the chronology. Finnian of Clonard died in 549, and Finnian of Magh Bile thirty years later. The first of these dates is far too early, the second also out of court, even though 588 for our Fionán's landfall can only be conjecture, based upon nothing more than sketchy archaeological and architectural evidence.

Where there is documentary support for any theory of title at this stage of Irish history, it is often confusing in its imprecision. The monk Adomnán of Iona, in his *Life of St Columba,* mentions one Finnbarr, with whom Columba studied as a young deacon in Ireland. Later in the same chapter, Finnbarr is rendered as Uinniau, and later still in the book, he becomes Finnio; and this monk has been identified by at least one scholar as the Fionán mac Airennain who is referred to in both the *Martyrology of Donegal* and the *Annals of the Four Masters,* the first source giving his feast day as 12 February, the second noting his death as late as 674. If he was indeed Adomnán's man, it would seem unlikely that he is also ours, and the chronology would also appear to rule him out, though that need not necessarily be so. He would clearly have become excessively old if he landed on the Irish skeilic in 588 and did not die till 674, but the centenarian sage is no rarity in early Christianity, or in any other faith.

Then there is Finan the Leper, whose festival occurs on 16 March; and Finan Cam, a pupil of Brendan's, whose feast day is held on 7 April. There is yet another Finan, an Irishman settled at one stage in the community on Iona who, in 651, succeeded Aidan, the founder and first abbot of Lindisfarne in Northumbria. Whether any of these correspond with two other names that have also been bandied about at various times—Finan of Loch-laoich and Finan of Loch-lein—is a question which has never been, and probably never will be, resolved.

Indisputably, however, there was a monastic foundation on a skeilic off the west coast of Ireland at some stage in this period; and a certain amount of circumstantial evidence on the mainland opposite the island indicates that St Fionán (as he would ultimately be remembered) was the man responsible. The search can be narrowed to County Kerry, to its Iveragh Peninsula in particular, where there are many associations with that name in a variety of its forms. There is a St Finan's Bay, for a start, and not far away what's left of a twelfth-century church with the same dedication, where, it was recorded in 1902, people suffering from ailments of a scrofulous nature still used to collect ferns growing from the outer walls, in the belief that if these were rubbed on the affected skin, they would alleviate the malady. At least one well in the district was also associated with Finan/Fionán. Some distance along this coast, on the flanks of Drung Hill, on the south side of Dingle Bay, there is a penitential station associated with Fionán, together with a well bearing his name, whose waters were highly regarded as a cure for diseases in cattle. And on the summit, there is to this day a monument which bears the name Leacht Fhionaín (Fionán's Altar). There, the saint is reputed to be buried; which, if true, would be unusual. Monastic founders generally left their bones with their communities.

3
Peregrinatio

The voyage made by Fionán and his companions, with its accep-
tance of a chance-directed outcome, was in the great Irish tradition of peregrinatio pro Dei amore, 'wandering for the love of God',
which was also closely linked with the notion of exile. In its most
abandoned form, this caused some monks literally to cast themselves
adrift at sea, happy to accept whatever the outcome might be. There
is at least one authenticated example of this, reported in *The Anglo-
Saxon Chronicle*, when the three Irishmen Dubslane, MacBeth and
Máelinmum in 891 'came to King Alfred in a boat without oars from
Ireland, from whence they had stolen away because for the love of
God they wished to be in exile, they did not care where. The boat in
which they came was worked from two and a half hides; they took
with them food for seven days. They came to land on the seventh
night in Cornwall, and went soon to King Alfred.' The same cast of
mind was also responsible for the migration of many Irish monks
overseas, from the sixth century onwards, of whom Columbanus
(Columban) was the most celebrated. Born in Leinster *c.* 540 (which
made him about twenty years younger than Columba/Columcille,

who came from Donegal) he left his monastery in Bangor, County Down, at the age of forty-five to wander across continental Europe with a dozen companions. Columban himself was to set up monasteries at Annegray, Luxeuil and Fontaine in the Vosges of Gaul, before crossing the Alps and making his greatest foundation at Bobbio, between Genoa and Piacenza, where he died in 615. His companions ventured elsewhere, and one of them started what became the celebrated monastery of St Gall in Switzerland.

There were many other such Irish peregrinations abroad. The ninth-century monk Dicuil and others settled in Iceland, where he composed a respected geography, *Liber de Mensura Orbis Terrae,* which drew extensively on the reports of a fellow Irishman, Fidelis, who had journeyed as far as the Holy Land. In Germany, the cathedral of Würzburg (perhaps the most wonderfully restored church in all Europe, put together again after the destructive barbarity of war) is to this day dedicated to the Irish saints Killian, Kolonat and Totnan, who were martyred nearby *c.* 689. Eventually, Irish scholar-monks, who were often skilled in the art of illuminated calligraphy as well, spanned continental Europe from the Atlantic seaboard of Brittany as far East as Kiev on the Dnieper. And all the continental monasteries founded by Irishmen like Columban and Gall became notable as centres of learning, as well as of extreme and rigorous devotion.

Yet the purpose of these wanderings abroad was in no sense missionary or the dissemination of scholarship: with the exception of pagan Iceland, where they made no attempt to proselytise, the Irish were, after all, remaking their lives in lands that were already Christian; and it was merely coincidence that these expatriates helped to keep the flame of western civilisation burning during the so-called Dark Ages, when the remains of the Roman Empire were overrun by Germanic and Slav peoples invading from the East. The intention of Columban, and others like him, was to achieve personal salvation in a union with God-in-Christ that they believed could best be achieved by distancing themselves as much as possible from all emotional and other ties that might bind them to their own people. Their writings and their biographies are strewn with phrases which indicate this cast of mind. One definition of peregrinatio is 'seeking the place of one's resurrection', while Adomnán refers to some of the monks setting sail from Iona because they were 'anxious to discover a desert in the pathless sea'.

Psychologically, they were better equipped than many others for such a bleak concept of their vocation, having been reared in a land where the fostering of children, physical alienation from their parents, was normal practice. From top to bottom of Irish society, children were separated from their families from infancy or shortly after it, until they were fourteen if a girl, seventeen if a boy, for reasons that may have been economic to start with but in time became cultural. The vast majority of these children after the arrival of Christianity were placed in the hands of priests or in religious houses, under the care of monks and nuns; and very many of them inevitably never returned to a secular life. An exception was Conall, son of Enda in Mayo, who was baptised by Patrick when a small boy and put in the care of Patrick's disciple Cethiacus; but returned to the secular state when Cethiacus died. More frequently, such was not the case. The *Book of Armagh* mentions the 'pueri Patricii' who were entrusted to Patrick as children and spent the rest of their lives in the Church. There may not have been a single founder of an Irish monastery, either at home or abroad, whose formation did not begin in this way. Columcille was fostered by the priest Cruithnecán before he passed to Finnian's monastery at Clonard. Columban was placed in the care of another holy man before he was put into a monastic school and then progressed to Bangor. Ciarán, founder of Clonmacnois, was the fosterling of a deacon named Diarmait. Brendan the Navigator, who started the neighbouring monastery at Clonfert, was one of many children who spent their early years under the care of St Ita, a woman whose religious instruction of the young became as proverbial as her asceticism.

It was generally understood in these circumstances that the child should be fostered in a monastery as far away as possible from everything that was familiar to it. That some who committed themselves to the religious vocation for life could nevertheless still be disturbed by the call of kinship, is clear from a conversation Columban had with a nun whose advice he sought when contemplating his own future. 'Fifteen years have elapsed,' she said, 'since I left my home and came to this place of pilgrimage. I chose Christ then as my leader, and I have never since looked back. Were it not, indeed, for the feebleness of my sex I would have crossed the sea and found a more suitable place of pilgrimage in a foreign land.' But eventually, the greater majority were able to dissociate themselves emotionally from all family ties. The monk Munnu received, at his anchor-hold in the Irish midlands, a message that his mother very badly wanted to see him at home in the North. He arranged to meet her halfway, and she arrived with other members of the family. Before they parted again, Munnu said this to them: "Take care never to come to me again, for if you do I will leave Ireland altogether and travel to regions beyond the sea.'

The tough and self-sufficient nature which developed as a result of these social and religious processes—and even more, perhaps, the quietly confident faith that underpinned it all—was perfectly expressed by an anonymous Irish religious in the ninth century:

> *All solitary in my little cell,*
> *With not a single soul as company;*
> *That would be a pilgrimage dear to me*
> *Before going to the meeting with death.*

For such men and women, for monks like Munnu, the peregrinatio meant venturing no further than some unfamiliar place in their native land; but a number, like Fionán and his companions, found their anchor-holds on the skeilics that are sprinkled around the Irish seas.

5
Pagans and Christians

A further characteristic of the early Church in Ireland was its continuity with the beliefs and understanding of creation that preceded Christianity there. Quite exceptionally in the history of the
religion, no one was martyred in Ireland—a land in which belligerence, from the earliest times, was never missing for long from any
level of life—with the coming of the faith and the conversion of the
native Celts. This has been persuasively attributed to the fact that
those first Christians and their descendants retained a striking affection for the local Celtic past, for its way of accounting for natural phe-
nomena, and saw no reason why they should not assimilate this
without compromising their new beliefs. They thus preserved much
of the pagan tradition in a modified form, and explained the origins
of their people in a series of typical legends. According to these, Ireland was populated by a series of migrations, the first of which was
made from Spain under the leadership of Partholón, an ill-fated expedition because it was followed by a plague which wiped everybody
out. Next, again from Spain, came Nemed mac Agnomain, who established a lasting colony. This was subsequently joined by the Fir

Bolg, a race of farmers from Greece, by the Fir Gálioin and the Fir Domnann from further west; all three names, significantly, corresponding with three historic Celtic tribes, the Belgae, the Gauls and the Dumnonii. Then came the Tuatha Dé Danann, 'the tribes of the goddess Danu', who conquered the Fir Bolg and another tribe, the Fomori, in two Battles of Moytura, the second of these under the leadership of Lugh, who was equally expert as a warrior, a carpenter, a smith, a musician, an athlete and a player of chess. Against the Fir Bolg, the Tuatha were led by the Dagdá, the 'lord of perfect knowledge', who was possessed of a magic cauldron, which never failed to produce enough food to feed the hungry, however numerous.

An early Christianised version of this mythology attempted to banish the Tuatha for obvious reasons, but so deeply rooted was the tradition that a compromise was reached in the revised accounts, in which the Irish appear never to have worshipped their old gods in any conventional sense, or made sacrifice to them, merely regarding them as supernatural beings with magical powers, with no deity claiming primacy over the others: the Dagdá was regarded as the greatest of them all, but he was content at times to play second fiddle to Lugh. Instead of total rejection, therefore, the early Christians channelled the Tuatha and their wizardry into a local underworld which could be visibly associated with the prehistoric burial mounds that were, and still are, such a feature of the Irish landscape. These became the sídhe (supernatural dwellings) of the Tuatha, where people would hold their tribal meetings, because they wished to take counsel of the dead as well as of the living. In this and in other ways—they were still held directly responsible for the fertility of the land—the Tuatha continued to be an active presence in the Celtic imagination, in spite of the new religion's imported beliefs. This accommodation between the pagan past and Christianity has been seen as the origin of the lingering Irish folk tradition of the little people, the fairies.

The Christians also adapted to the old Celtic calendar, which had four great focal points of ritual. The traditional new year began with

Samhain, at the beginning of November (corresponding to All Souls' Day), an uneasy time when the boundary between this world and the next seemed fragile, when spirits could cross over and the living could be drawn away; and when masked children traditionally collected overdue offerings on behalf of the Tuatha, in gratitude for a good harvest. Halfway through winter came Imbolc, when there were fertility rites associated with sheep, which eventually became the Christian feast of St Brigid on 1 February. Brigid (or Brigantia) in the Celtic mythology was a daughter of the Dagdá, while in the Irish Christian tradition, she was midwife of the Virgin Mary; and a further blurring occurred after 523 when the abbess Brigid, 'the Mary of the Gael', founder of a dual community of monks and nuns in Kildare, died there and became associated with the February feast. When 'the lighted half of the year' began at the beginning of May, a second fertility festival was Beltaine, whose spring optimism could without difficulty be celebrated as a palimpsest of Easter. Finally, there was the August festival of Lughnasadh, associated with Lugh and with harvest in the old tradition, which also required merely the transposition of a deity after the Christians arrived.

Nor was the Christian doctrine of the Trinity any obstacle to a people who held the trefoil sacred, and who were accustomed to grouping their deities in triads—Mórrígan, Macha and Badb, goddesses associated with war and death; Brigantia/Brigid and her two sisters, who were the patrons, respectively, of poets, smiths and laws. The continuity between the druidical past and the priestly future was such that Irish Christians frequently anointed themselves so that the Devil could not come to grips with them. As the Romans had noted more than once in pre-Christian times, all Celtic warriors habitually oiled their bodies to prevent their enemies seizing them.

A result of these adaptations, this interplay between the two systems of belief, was that the Celts—after, it seems, one false start earlier on—accepted the new faith without any marked sense of imposition, from the hands of Patrick in the fifth century. And, as Christians, the Irish were never required by the Church that Patrick began, for as long as it remained a distinctively Celtic Church, to abandon one of their most profound instincts, which was to see God in everything, and not simply as a supervising deity detached from the natural world except on the one occasion when Christ came to earth and lived among humankind. This was something very close to pantheism, separated from it, maybe, only by an intense devotion to the person of Christ, a fixation with joining Him in heaven after death. If there is a key text in this amiable fusion of two traditions, then it is perhaps something that originated in the oral culture of Ireland's most ancient past. The old stories say that there was a poet, Amargin, who led the Tuatha Dé Danann when they arrived in Ireland. To him is attributed some Druidic verse which no less an authority than Douglas Hyde— himself a poet, a literary scholar and the first President of Ireland (1938–45)—believed might be the oldest surviving lines in any European vernacular outside Greece. This was Dr Hyde's translation:

> *I am the wind which breathes upon the sea,*
> *I am the wave of the ocean,*
> *I am the murmur of the billows,*
> *I am the ox of the seven combats,*
> *I am the vulture upon the rock,*
> *I am the fairest of the plants,*
> *I am the wild boar in valour,*
> *I am a salmon in the water,*
> *I am a lake in the plain,*
> *I am a word of knowledge,*
> *I am the point of the lance in battle,*
> *I am the God who creates in the head, the fire.*
> *Who is it who throws light into the meeting on the mountain?*
> *Who announces the age of the moon (if not I)?*
> *Who reaches the place where the sun couches (if not I)?*

5
St Patrick

The false start of Christianity in Ireland is an established fact, though there is no certain knowledge when the faith was first offered to the local Celts before Patrick's time. The most unlikely premise is that an Irish warrior in the service of Rome, named Althus, was present at the Crucifixion, beheld the miraculous events which happened afterwards, and subsequently hastened home to tell his fellow countrymen what he had seen and heard. Another legend maintains not only that Irishmen travelled to Egypt to converse with the Desert Fathers in the early days of Christianity there, but that Coptic monks in turn journeyed to Ireland, where seven of them are supposed to be buried. There is more substantial reason to believe that Christianity arrived not later than the fourth century—that is, maybe as much as a hundred years before Patrick—as a result of the commerce between Ireland and Gaul, which was well established by then, or between Ireland and Galicia, in northwest Spain, which had flourished since before the time of Christ; for in both Gaul and Galicia there were known Christian communities by 325. The first documented account shows that Pope Celestine I made his deacon Palla-

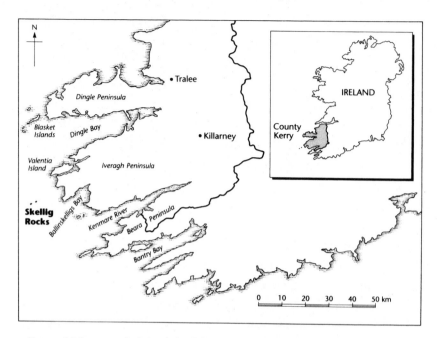

dius a bishop and despatched him to Ireland in 431. Sailing from
Gaul, Palladius landed and preached at Wicklow, but left the country
shortly afterwards for some unspecified reason, and died a little later
in Britannia. A memorandum written by a papal official asserts that
Palladius was sent on his mission 'as the first bishop to the Irish believing in Christ'; and there is further evidence that Christians were already settled in Munster and Leinster then. Local tradition included
a number of holy men before Palladius: Declan of Ardmore, Ilbe of
Emly, Ciarán of Saighir, Abbán of Moyarney, Ibar of Beg-Eire.

Hard on the heels of Palladius came the man by whom, in the popular imagination ever since, Ireland was truly converted to Christianity. He was born about 385–90 at a place he calls Bannaventa Berniae,
which is thought to be, exasperatingly, 'somewhere between the Rivers
Clyde and Severn', with the area of Scotland near modern Dumbarton
attracting more nominations than anywhere else. He was Patricius, the
Romano–British son of Calpornius, who was a decurion, a municipal
official, and a deacon in the local Church, in which the grandfather

Potitus was a presbyter. When Patrick was almost sixteen, he and his sister were abducted by Irish raiders commanded by the notorious chieftain Niall Naoi Ghiallach (Niall of the Nine Hostages), who had come over the water in a fleet of curachs looking for plunder and slaves. The girl finished up in Connacht, Patrick in County Antrim, not far from Ballymena, where he was put to work for the chief Miliucc (Milchu) as a shepherd and swineherd: a life of great hardship in which, as may be imagined, he often went hungry and was never adequately clad. But in his early twenties he had a vision, in which he was told that he would soon be returning to his own land, and that a ship was awaiting him some distance away. The distance was about two hundred miles to the south, but he walked every inch of it, sailed from some place near Wexford and crossed over into Gaul.

There are contradictory accounts of what happened next, the only certainty being that he was going in the wrong direction to be reunited with his family. In one version he travels southeast to Milan and thence to the Tyrrhenian Sea; in another he goes straight down Gaul until he comes to the Mediterranean. But it is agreed that eventually Patrick entered the famous island monastery of Lérins, off the coast of Provence, which had been founded by the Romano–Gallic convert Honoratus not later than 410. We do not know how long he stayed there (one estimate even puts it at twenty years) but in time the pull of family became irresistible and he left for Britannia where, presently, the decisive moment in his life occurred. In his own words,

I saw in the middle of the night a man whose name was Victoricus, coming as it were from Ireland with countless letters. And he gave me one of them, and I read the beginning of the letter, which was entitled The Voice of the Irish; and while I was reading aloud the beginning of the letter I thought that at that very moment I heard the voice of them who lived beside the wood of Foclut, which is nigh unto the Western Sea. And thus they cried as with one mouth, 'We beseech thee, holy youth, to come hither and walk once more among us.'

At this stage, Patrick had only been ordained deacon, but something weightier was required for the task in hand. He returned therefore to Gaul, to study under Bishop Germanus in Auxerre. There is reason to suppose that Patrick, and not Palladius, was the first choice to be consecrated bishop in order to lead an Irish mission but that there was opposition to this by clergy in Auxerre on two grounds, one of which was Patrick's limited education. The other has tantalised posterity ever since. Patrick had a guilty secret from the days of his youth—'what I had done in my boyhood on one day, more precisely in one hour, because I had not yet prevailing power'—which he had only ever revealed to one person, who now chose to disclose it in Auxerre. One speculation is that he had killed a slave before his abduction, though it seems at least as likely that his guilt was caused by some sexual transgression. Whatever the secret was, its disclosure appears to have been enough to delay Patrick's consecration until Palladius's death, at which point there really was no other suitable candidate for the Irish mission.

In 432, he landed with a few companions where Wicklow town now is, before sailing on up the east coast, stopping here and there to preach and to accept conversions. He came to the Hill of Slane and there lit a paschal fire in order to keep Easter, not knowing that he was within sight of Tara, traditional seat of royalty since the time of the Fir Bolg, and now occupied by Laoghaire, eldest son of the Niall who had kidnapped the young Briton. Laoghaire was preparing to hold the feast of Beltaine, and there was an inviolable rule that no other fire which was visible from Tara could be lit that night. Druids urged the king to act and, with nine chariots and two magicians, he made for the Hill of Slane and confrontation with the Christian upstart. The Druids, of course, had the worst of it, one of them being whirled into the air and fatally dashed against a rock; whereupon darkness covered the earth, which trembled, and caused the king's men to fight each other in the murky confusion, though Laoghaire tried to kill Patrick and his disciples—who at once turned into stags and leaped away to safety. Laoghaire came to his senses after this, was baptised, and gave Patrick safe conduct through Ireland, his own son Erc being converted at the same time, eventually becoming the first Bishop of Slane. This episode also includes the story of Patrick's supposed dissertation on the Trinity, using a shamrock to illustrate his theology.

The saint—all Celtic holy men and women bore this title, often enough in their own lifetimes, though very few were ever formally canonised—proceeded through the kingdom of Meath, and then took the high road to the west, accompanied by a growing retinue of converts and clergy whom he installed in churches en route, including the son of his old captor Miliucc (one can picture Patrick here as an early Irish version of Mahatma Gandhi, when he strode across India to the sea at Dandi in order to defy the salt tax, with scores and then hundreds running to keep up with Gandhiji; many of whom, interestingly, identified with Christ and His disciples marching to Jerusalem, to the extent that they read the New Testament on their way to the Arabian Sea). Patrick stayed seven years in Connacht, and it was there, in answer to questions, that he made his pronouncement on the nature of God that was not so far from the existing Celtic concept of divinity.

'Our God,' he said,

is the God of all men; the God of heaven and earth, of the seas and the rivers. The God of the sun, the moon and all stars. The God of the high mountains and of the lowly valleys. The God who is above heaven and in heaven and under heaven. He hath an habitation in the heaven and in the earth and sea and all that are therein. He inspireth all things. He quickeneth all things. He is over all things. He sustaineth all things. He giveth light to the sun. He hath made springs in a dry ground; and dry islands in the sea. And hath appointed the stars to serve the greater Light. He hath a Son, co-equal and co-eternal with Himself. The Son is not younger than the Father, nor is the Father older than the Son. And the Holy Ghost breathes in them. The Father, the Son and the Holy Ghost are not divided.

It was while in Connacht that the saint is supposed to have climbed the Hill of the Eagle (Croach-Aigli, which eventually became Croagh Patrick, overlooking Westport Bay), enticing all the toads and serpents and anything else venomous in the land, which he despatched into the sea with the help of his miraculous staff; a gift from Christ Himself, it was said, during Patrick's time beside the Mediterranean. From Connacht he went to Ulster, preached in Donegal and established the see of Amargh, a century and a half before there was a bishop of Canterbury. Then he went south to Munster, where idols fell at his coming to Cashel, before he returned to the region near Strangford Lough in the North, where he died in the monastery he had founded at Saul. The year was 461, one of only two dates in the entire story of Patrick which are regarded as historically reliable; the other is 432. The monks of Saul yoked two untamed oxen to the cart which bore Patrick's body, and left them to make their own way where they would without guidance. When they stopped, Patrick was buried; and over his grave was later built Downpatrick Cathedral.

Great caution is necessary when contemplating the life of Patrick, whose own writings are vague when it comes to historical detail, whose first biography by another hand was not written until the end of the seventh century, and who had become the subject of innumerable fantasies long before that; so that what we think we know about him is a mixture of fact, embellishment and wishful thinking. It is often impossible to distinguish between one and another. But whatever the truth of it, he was, without question, a towering figure in the history of his adopted people. His own compositions, in Latin, were the very first words written down in Ireland, using the Roman alphabet rather than the more primitive ogham script, which was mostly inscribed upon stone. They contain no miraculous stories.

6

Brendan the Navigator

The life of Brendan the Navigator is, if anything, even more improbable than that of Patrick, according to the received accounts. It could scarcely be otherwise when there are said to be 120 or more different versions of his most famous exploit, written between two and five hundred years after his death. All of them, according to Douglas Hyde, were 'probably founded upon some actual tradition, which was built up into a wildly fictitious romance in the seventh or eighth centuries after the fashion dear to the hearts of Hibernian authors of that age'. An indication of the uncertainty surrounding these editions of *Navigatio Sancti Brendani Abbatis* is that the given number of crew who accompanied him on his greatest voyage varies, in one text and another, from no fewer than fourteen up to as many as sixty souls. But the possibility that a sixth-century Irish monk did make a remarkable Atlantic voyage had certainly not escaped the attention of Arab geographers in twelfth-century Spain. Columbus was also aware of the *Navigatio,* under the tutelage of that vigilant genius Henry the Navigator, who was plotting Portugal's great maritime destiny in the fifteenth century on his clifftop at Sagres.

Brendan was born *c.* 484 at Annagh, a marshy place not far from Tralee, to parents who shared the royal blood of Ui Niall, and who had probably been baptised by the Bishop Erc whom Patrick had converted at the court of Laoghaire. Varying the tradition of fosterage enshrined in the ancient Brehon law of the Irish to suit their own circumstances, the local Christians had by this time determined, a generation after Patrick, that every first-born child of a properly married couple belonged to the Church. And so, at the age of one, Brendan was handed over to Ita and her convent school of Ceall Ide (now Killeady, in Limerick), where he stayed until he was six, when he was returned to Kerry and placed as an oblatus under the care of another Erc, the Bishop of Altaighe Caille, in order to continue his education. He would certainly have been taught Latin, possibly some Hebrew as well; we can be almost sure he knew no Greek, for there is no evidence that it was taught in any of the early Irish monasteries. At some stage, Brendan became one of the peregrini who were beginning to wander the land in search of their spiritual vocation, but he returned to Erc at the age of twenty-six and was priested by him.

He acquired followers and with them repaired to the slopes of a mountain on the Dingle Peninsula (it was known as Dadche then; it is named after Brendan now) overlooking a bay, on whose shores it was decided to build a boat. It seems that he had become fixated with the idea of discovering the Promised Land of the Saints—a Christian variant of the ancient and universal legend involving the quest for a beguiling island of Eternal Youth, which the Irish had always known as Tir na nOg. The difficulty for those who have tried to interpret what is recorded after this in the *Navigatio* is that the oral accounts of Brendan's expedition which preceded the first written version by anything up to a century, almost certainly became entangled in the imagination of the scribes with other examples of the Celtic iomramh, the voyage story, which had a similar and in some cases slightly older pedigree. One of these was the pagan Iomramh Bhrain (the voyage of Bran), which begins with a woman serenading the hero about the sensuous delights of the Other World, causing him to go looking for it with twenty-seven companions. Another was the Iomramh Curaig Maíle Dúin (the voyage of Mael Dúin's boat) and belongs to the Christian era, with a nun cast as mother to the hero, who sets off across the sea to find his father's murderer.

From a jumble of sources, then, the following outline of Brendan's great voyage emerges. He and his companions are about to embark when three other monks rush up and beg to be included in the crew. Unwilling at first, Brendan is finally persuaded, but warns them that two of them will die hideously, that none of the three will return; but this does not cool their ardour. The voyagers sail off westwards but after fifteen days are first becalmed and then blown onto an island, where food has been set out as if awaiting them, although they never see any creature but a stray dog. One of the three latecomers tries to steal a silver bridle he finds there, but a demon leaps out of his chest and he drops dead. His companions bury him at sea and the curach sails on until it reaches the Island of Sheep, with endless and handsome white flocks, where the monks spend part of Holy Week. An islander feeds them here and foretells that they will spend Easter on a nearby island before going to the Paradise of Birds, where they will remain until the eighth day of Pentecost. He gives them mutton to take with them and they are about to start cooking it on the beach of their next stop, when the island begins to shake and burn, so they put to sea again in alarm, watching the island drifting away in flames. Some time later they do gain the Paradise and, at the head of a long stream, find a tree smothered in white birds, one of which flies to Brendan and explains to him that it and the other birds are the spirits of men, and that the search for the Promised Land will take seven years.

The voyage continues for three more months, after which, exhausted and almost without food or drink, they sight land again. On reaching the shore with difficulty, they are met by a white-haired old man who leads them to a monastery, where silent monks embrace them, and the abbot washes the travellers' feet. This is the Community of St Ailbe, a place where no cooked food is ever taken, only sweet roots and white bread, where a fiery arrow flies through the window to light the lamps in church, and where the monks have dwelt for eighty years without growing older or becoming sick, hearing no human voice, communicating only by signs. Brendan and his men spend Christmas there, proceed on their way during Epiphany, and keep sailing until Lent, when their victuals once more are running out. They find another island, where there is water and edible plants as well as fish, but when the water puts some of the monks to sleep for days, Brendan decides they must leave. Three days later they are becalmed and Brendan orders his men to ship their oars and drift under the providence of God, Who will decide which way their craft should go next. A wind arises and blows them to the Island of Sheep again, where once more they are warmly received before calling again at the Paradise of Birds. On the way there they encounter a whale, which reappears more than once before the voyage is done, and which allows them to land on its back. After another forty days a less tractable monster begins to pursue the boat, spouting foam and closing in at great speed to smash the curach and devour its crew; but another monster (this one breathing fire) heads it off and cuts the first beast into three before swimming away again.

The saga continues with other islands interrupting the long passage at sea. One of these is covered in purple-and-white fruit, boasts three different choirs (boys in white, young men in blue, elders in purple) who sing for the voyagers and supply them with the fruit, each piece of which tastes of honey and feeds one man for twelve days. As the monks prepare to leave, the second of the late additions to their crew decides to stay behind with the choirs. Some days later a giant bird flies over the boat and drops grapes the size of apples, subsequently driving off a flying gryphon which is about to tear the curach apart with its talons. Another Christmas is spent with St Ailbe's Community, further visits are made to the Island of Sheep and the Paradise of Birds. The wanderers come across a huge crystal pillar rising from the sea, so high that they cannot see the top, then two islands which emit flames and smoke. At the first they hear the sounds of bellows and hammering, and men fling lumps of burning slag at them, which miss the boat, but produce a great stench as they fly overhead and make the sea boil where they fall; from which Brendan concludes that he and his crew have reached the edge of Hell. At the second island, dominated by a mountain, they are driven aground by the wind, and the third of the ill-fated monks leaps from the boat, saying he is powerless to do otherwise. They watch as demons carry him off and set him on fire, whereupon a favourable wind blows the curach out to sea again and, when the mariners look back, they see that the island has become a colossal pyre.

Their next encounter is with a man sitting on a rock, who turns out to be Judas, spared by the Lord from the flames of Hell on holy days, but at those times condemned to exposure amidst the breaking waves. Demons pursue the curach as it sails away, then snatch up Judas and carry him off to his normal fate. Three days later, another island appears and on it Brendan finds an anchorite who is 140 years old and who has lived there for ninety years on fish brought to him every third day by an otter. This ancient advises Brendan to provision well, because he faces a forty-day voyage back to the Island of Sheep and the Paradise of Birds, after which another forty-day journey will take him to the Promised Land of Saints, where he will be allowed to stay for that same length of time, after which God will bring him safely back home.

Back they go to the Island of Sheep, where they are met by their old friend the man in charge, who joins them for the climax of the epic. First the curach and its complement are carried over to the Paradise of Birds on the back of the amiable whale. Then they sail to the east, with their friend in the bows, where he can direct the steersman through a thick fog that has befallen them; which, he says, perpetually encircles the land that Brendan has sought these seven years past. Presently a great light shines through the mist and the boat comes to a land full of fruit trees, which the monks explore without ever reaching the end of it. But they find a wide river where a young man embraces them individually, calling each monk by name, and explains that God has delayed their arrival so long in order that they shall know the secrets of His ocean. He tells them that the land will be revealed to Brendan's successors if ever the Christians of Ireland are persecuted; and that now, after revictualling and collecting precious stones, they should return whence they have come. And so Brendan's great voyage comes to an end, with a return passage through the fog until Ireland is on the bow.

The allegory and the mystic purposes of the voyage are the stuff of mythology and are of no more than intrinsic interest: what has fascinated sceptical scholars much more ever since are the navigational details. Many attempts have been made to identify the landfalls mentioned in the *Navigatio,* and some conclusions are more plausible than others. The Island of Sheep may very well be in the Faroes (the Danish word for sheep being faar), the stop before that possibly St Kilda. Iceland is almost certainly the origin of the fiery islands with their slag-throwing inhabitants, and we may well think that the Community of St Ailbe was located in the Shetlands, where Irish peregrini are known to have been settled in time to be included in the *Navigatio.* The crystal pillar rising from the sea is obviously an iceberg. But other speculations are much more ambitious than any of these—the possibility that Brendan not only reached Newfoundland, but that he sailed on to the Bahamas, and round the Caribbean until he came to a Promised Land which was Florida. On the other hand, it has been demonstrated in modern times that a trans-Atlantic crossing in a similar craft to the saint's is perfectly possible: with a much smaller crew than Brendan had at his disposal, the explorer Tim Severin sailed, insofar as it could be determined, the legendary route in 1976–7. And before dismissing this sixth-century feat out of hand, it is worth remembering that, sometime before 300 BC, Pytheas of Massilia (Marseilles), having already become proficient enough in astral calculations to be no more than fourteen miles in error when estimating the latitude of his home port, sailed from the Mediterranean and returned safely after evidently circumnavigating the British Isles and probably getting as far north as Iceland or Norway.

The burial of the monk at sea in the *Navigatio* is the first recorded instance of such an interment in Northern Europe, but little else about Brendan's seafaring is susceptible to proof. There have been suggestions that the account amalgamates two or more voyages, presenting them as a single epic; that one journey into the Atlantic was made in a curach with ox-hides forming its hull, another in a wooden boat with an iron anchor; and that there were other ventures which took Brendan round the Northern Isles of Scotland, down to Wales, across to Brittany (landing on an island outside St Malo), even as far as the Canary Isles (there is a statue to him in Tenerife). There is also a fable that he visited Egypt and Palestine, bathing in the Jordan before going home. We are on safer ground when he is back in Ireland, an ageing monastic seadog perhaps, but still with many years to live. He died in 577 at Annaghdown, County Galway, where his sister Brigh was the superior of a convent he had founded, but he was buried close to the River Sinann (Shannon) at Clonfert, his own monastery, the one in which he reared Fionán.

Brendan had established Clonfert almost twenty years before his death, in a meadow that was flanked on one side by the river, on the other by forest; it was not far downstream of St Ciarán's monastery of Clonmacnois which, though only a little older, already had a towering reputation for scholarship, which would scarcely be equalled anywhere in the Irish monastic tradition. Brendan was in his nineties when he died, and seemed troubled as he approached the end. Brigh asked him what he had to fear in death. He replied: 'I fear going alone, for the journey is dark. I fear the unknown, the presence of the King, the sentence of the Judge.'

6

Irish Monasticism

The norm in Irish monasteries at this time was utterly outside the convention that would be established later by the great continental orders, of whom the Benedictines and the Cistercians are the most familiar. No Celtic monastery consisted of a community under one roof. Instead, each was a settlement of many separate buildings, adjacent to or surrounding a church, sometimes more than one: Clonmacnois is said to have had twelve, though the fact that they were usually referred to as oratories perhaps gives a better idea of their size than does the word church. Most numerous among the buildings were the cells of the monks. Others served as workshops, infirmaries, kitchens, refectories, schoolrooms, guest-houses; and, at this stage in their history, the majority were wooden or made of wattle and thatch, stone being used only where wood was unobtainable, with the entire monastery encircled by a ditch and the earthen rampart of its rath. The abbot would have a hut of his own, and so did some of the oldest monks in each community, as a concession to their growing infirmity; but the majority of the brethren seems to have shared huts in twos and threes or even more.

It is recorded that by the sixth century there were monastic settlements both great and small throughout Ireland, though these terms are
relative and in some cases the figure given for a monastery's manpower
has been misleading. Traditional accounts of Brendan claim that, at his
death, Clonfert housed three thousand monks and students, who had
been attracted by the saint's scholarship and piety as well as by his ad-
venturing; Clonard and Clonmacnois were reputed to be in the same
category. Such figures are almost certainly gross exaggeration, in the
same way that repeated references in the *Navigatio* to stages in Brendan's voyage lasting forty days are no more than a vivid way of saying
that they took a long time. It is now thought that, at the most, the
largest of the establishments would run to no more than a couple of
hundred men and boys; the figure could, however, be greatly swollen
if a monastery's dependent houses were taken into account, and these
were frequently prolific. An abbot often had under his obedience a
group of monasteries, single churches and separate hermitages, which
formed his parochia, a word that gradually acquired the modern understanding of 'diocese'; but at this period an abbot had much greater
authority in the area of his jurisdiction than any bishop. When he authorised one of his monks to detach himself from the mother house
and found a new establishment of his own, it followed that the monk's
obedience ceased and he himself became the head of another parochia.

From the outset of Irish monasticism, the head of a community was
known as Abb, Abba or Abbas (from the word for 'father' in Aramaic,
which was Christ's language) and his elder brethren were the seniores (sruthi in Irish), who exercised a considerable authority on their
own account over the younger monks. The oblatus, who was installed
in a monastery in childhood with the expectation that he would remain there for life, was uncommon until much later in Irish history;
and the adult conversus never did form a significant part of the
monastic population. Most monks entered between the ages of fifteen
and seventeen, having already been taught letters and the rudiments
of the faith from some pious individual, or from some other commu-

nity on whom they had been fostered as children. They would first of all be received as guests and questioned searchingly about their motives before being admitted. And although there was no master of novices, with a supervisory role over all newcomers, such as would eventually become a regular feature throughout Christendom, one of the seniores would be told to keep an eye on the young entrant, who was generally expected to learn by imitation. A striking feature of these recruitments is that they included very few young men of lowly birth. Many were the sons of chiefs and kings; the rest came mostly from families which, in today's terms, were of the professional middle-class. Their fathers would be superior craftsmen, men of some standing in their communities.

The rigour of the life these novices had embraced may be judged from the Rule that Columban took with him from Ireland to continental Europe. 'Let the monk,' he says,

live under the authority of one paternal ruler and in company with many brethren, so that from one he may learn humility, from another patience, let one teach him silence, another gentleness of manner. He must not do what he wants to do; he must eat what is placed before him; he must have nothing but what he has received; he must perform the task assigned to him, and show submission to one whom he does not like. He should be tired out before he goes to rest and be half-asleep while still on his feet, and be compelled to get up before his need for sleep has been satisfied. When he is insulted he must listen without a word. He must fear his superior as a lord, but love him as a father, and must believe that whatever he orders is for his good. The judgement of anybody placed over him he must never criticise, for it is his duty to obey and to do what he is told . . .

Novices were frequently given menial tasks such as washing the feet of their brethren, to induce an appropriate humility. So little was his regime to the taste of some continental monks who came under Columban's obedience that eventually he was obliged to defend himself to Pope Gregory; but to Irish monks, such strictness was unexceptional.

6

Book Illustration

The monasteries of the sixth and seventh centuries were wholly responsible for Ireland's reputation as the most learned country in western Europe, when continental scholarship was all but extinguished by wave after wave of barbarian invasion. This was terrible enough, but in some senses more grievous still was the Church of Rome's contempt for learning in this period, to the extent that Gregory the Great knew no Greek and was deeply opposed to anything being taught other than Holy Writ and its elucidation. The Irish monastic schools were therefore not only bursting with native youths eager for the education they provided, but also attracted students from abroad, anxious to avail themselves of an opportunity which was denied them at home: Clonmacnois soon developed a particularly strong relationship with intellectual refugees from Gaul, one of whom would become the Merovingian King Dagobert in 622.

The teaching varied in some particulars from monastery to monastery, but it was almost entirely in Latin with a certain amount of Greek eventually as a supplement, and the first thing every student had to do was to learn the full Psalter off by heart, before taking instruction in Holy Scripture generally. Beyond that, which took precedence without exception, Irish teaching was notable for its excursions into secular topics as well as basic Christian works. It is said that at Clonfert, Fionán and the other young monks would have learned mathematics, history, law, medicine, poetry, music, dialectics, astronomy, navigation and chess, as well as theology. Classical authors taught there included Persius, Virgil, Horace, Sallust, Ovid, Juvenal, the Christian poets Juvencus, Prudentius and Ausonius; also studied were works by Athanasius, Cassian, Jerome, Origen, Augustine, Cyprian, Gregory (the Great and narrow, no less), and the paschal cycles of Anatolius, Theophilus, Cyrillus, Morinus, Victorius, and Pachomius. Even allowing for the exaggerations that seem to be built everywhere into the Brendan hagiography, Clonfert must have been an impressive intellectual forcing house. And it was by no means unique.

Each of the principal monasteries contained its scriptorium, in which teams of monks copied the necessary texts in the intervals between their devotions. Until Gutenberg invented moveable type and the printing press at Mainz in 1440, the only way to create a library, or a book for everyday use, was by laboriously transcribing a new copy from an existing volume; and monasteries throughout Europe were the sole source of supply for centuries. The reason Irish monks were required to learn the Psalter by heart was because it was the only text each of them needed to be familiar with, in whole or in part, every day of their lives: the Gospels were not chanted chorally, as were the psalms, but were read or recited by an individual; and the missals were only necessary to the priests, who for a long time would remain a very small proportion of any community, the vast majority of monks being lay brothers. The creation of Psalters for scores of readers would obviously have led to enormous production problems in any scriptorium, to the detriment of its other work; and a people reared in a great folk tradition of oral storytelling, as the Irish were, would not find the prospect of learning 150 psalms by rote nearly as daunting as would Europeans of the late twentieth century.

Before the calligraphy and the artwork, it was necessary for the monastery to produce its own materials. The cost of making vellum pages from the skins of calves or lambs would have been prohibitive to all but the largest monasteries, when a volume of 250 leaves might require half that number of skins, each more than two feet by fifteen inches in size. So the smaller houses instead became accustomed to inscribing waxed wooden tablets, which were then pierced along one edge so that they could be crudely bound together with leather thongs; which is similar to the earliest form of book produced by Buddhists in India. An example found in County Antrim in 1914 had six leaves of yew, each a quarter of an inch thick, three inches broad and just over eight inches long, with tie holes for the thongs and a leather carrying strap. Each leaf, apart from the two outside faces, was slightly hollow on both sides except in its margins, to accommodate the beeswax. The leaves were incised with Psalms 30, 31 and perhaps 32. This book is thought to have been made in the seventh century. Such artefacts, however, were not the source of the great Celtic reputation for unsurpassed works of art in manuscript form. That was made in the scriptoria of monasteries like Clonmacnois and Clonard and Clonfert, on Iona, in Northumbria, and in other places which were under the Irish influence; one of the finest examples is the Würzburg Codex, an Irish vernacular manuscript, dated *c.* 700.

The writing materials everywhere consisted of those Fionán had to produce for himself at Clonfert, though the white of eggs was used to bind the colours to the page as often as fish oil, and lamp black rivalled chimney soot as a basis for making ink. Other substances which were pulverised or strained in order to extract dyes included gold (used very sparingly, for obvious reasons), white and red lead, malachite (for yellow) and folium (a schistose rock which yielded a variety of shades from blue to pink and purple). There were at least two animal sources of colour, one being the common dog whelk, which was found on every Irish and British coast and contained a liquid that, on exposure to the sun, turned green, then two different shades of purple. The other was an insect, kermes ilicis, whose female creates warts on a Mediterranean species of evergreen oak that produce a vividly carmine red when crushed. The costly blue mineral that Fionán coveted was, of course, lapis lazuli, whose only known source of supply was Badakhshan in Central Asia. And if that seems an improbable longing in a sixth-century Irish monk, it should be remembered that trade had been coming from China to the Mediterranean and beyond for several hundred years by then: the dried roots of rhubarb, much esteemed for their laxative properties, are known to have been conveyed to Europe by caravans travelling the Silk Route as early as 114 BC.

The volumes produced by the monks varied in size, some of them small enough to fit into a decent pocket. But heftier works like the *Book of Kells,* the *Book of Durrow,* the *Book of Dimma,* the *Lindisfarne Gospels,* the *Durham Gospels* and the *MacRegol Gospels,* would have been difficult to handle without a lectern, often having heavily embossed covers of leather and metal to protect pages that, in the typical case of Kells, were thirteen by nine and a half inches apiece. The first of the illustrated Scriptures appeared in Christendom about 400, though another four hundred years or so would elapse before a complete Bible, as we understand the word, was produced. Before then, monastic scriptoria turned out separate volumes of each Gospel, and for a while they produced nothing else from the New Testament except the book of Revelation: the twenty-seven parts of the New Testament were never mentioned before 367, when Athanasius referred to them. Similarly, for the first monastic scribes, the Old Testament was only represented by the five books of the Pentateuch—Genesis, Exodus, Leviticus, Numbers and Deuteronomy—which supplied a narrative from the creation to the death of Moses, just before the Israelites reached the Promised Land.

In the Celtic volumes there were, by and large, three different kinds of page. One sort carried the text, usually in a single wide column, but sometimes in two or even more narrower ones; it would include an ornate initial letter at the start of each page or paragraph or even line;

and there were occasional pictorial embellishments, such as the fish that Fionán indulged himself with to decorate the letters IHS as an abbreviation for the name of Christ. The script itself might be in the form known technically as half-uncial, which was a variant of formal Roman calligraphy, though in the *Book of Kells* the development of the cursive insular majuscule and minuscule styles can be seen in later pages, an Irish contribution widely copied elsewhere because letters were joined, and this technique was therefore time-saving. The contents of these pages obviously varied to some degree, from one monastery to another: the Irish held mostly to the Old Latin version of the Gospels, which went back to the second century and may have been composed in North Africa, whereas Lindisfarne and others preferred St Jerome's Vulgate, a revision he had made in the fourth century from the original Hebrew and Aramaic of the Old Testament as well as the Greek of the New, and which quickly became the standard version in the Western Church, except on its Celtic fringe. But both Kells and Durrow are peculiar—and raise interesting questions about their origins—because they mix the two translations to some extent. No volume limited itself to the scriptural text. There were also Canon Tables, enabling the reader to find his way through the Gospels, which were presented in numbered sections and not in the chapters and verses to which we are accustomed (chapters were not introduced until the thirteenth century; verses had to wait until the sixteenth). There were Breves Causae, which summarised each Gospel before its opening words; and the Argumentum, which were anecdotes about each Evangelist, preceding his particular text; and 'glosses', which were explanations of Hebrew names that were about to crop up.

Other than the textual pages, these volumes contain purely ornamental carpet pages, so-called because their largely abstract grid designs somewhat resemble the patterns extensively used by oriental carpet-makers, though invariably the pages also incorporate the cross. Most captivating of all to the inexpert eye are the pictorially decorated pages, which may be filled by a single fantastically complex initial, or which may exhibit portraits, caricatures, fantasies of human beings and other creatures. A famous example in the *Book of Kells* is the page which contains fantasised symbols representing the Four Evangelists: a man for Matthew, a lion for Mark, a calf for Luke, an eagle for John. And in these highly stylised examples of Celtic art, in the zoomorphic ornaments which represent writhing mammals, birds and reptiles, as well as the anthropomorphic depictions of the human form, some specialists have seen a connection with the art of cultures that extend far beyond the Celtic origins in Mitteleuropa, but can perhaps be traced to the Assyrians, the Persians, the Chinese, the Turkmen of the Central Asian plateau.

Wherever this imagination began, its development in the scriptoria of the Irish and other Celtic monasteries was nothing less than the evidence of genius. No work of art repays careful study more than the page of a Celtic manuscript, which is so copious and intricately related in its detail that you have to look at it long before your eyes become, as it were, accustomed to its light. Only then do you begin to notice the half-concealed creatures writhing around the margins, or the fact that St Matthew appears to have been given two right feet, or the colour of human eyes, which much more often than not are grey or brown, while the hair of the saints and angels is invariably in the range of yellow to red; or even prepare to try deciphering the word 'Initium', which occupies a whole page at the beginning of St Mark. You have to take your time with such frequently bewildering richness as this. (It may have become expedient, but it is nevertheless monstrous that the only way to see the *Book of Kells* in Trinity College, Dublin, nowadays is in an endless column of industrial tourists, who are kept moving by a series of intimidating functionaries, at approximately the same speed required of visitors to Lenin's mausoleum in Moscow). The genius is all the more striking because none of these books was the product of a single hand. Three different monks, at least, are thought to have been responsible for the *Book of Kells;* similar analysis would doubtless tell the same tale elsewhere. And such a painstaking, steadfast genius it was. There is an initial page in the *Lindisfarne Gospels,* with the scribe's imagination let loose on the opening words of St Luke, full of grotesque creatures, tendrils and ornaments as well as beautifully penned uncials; and with all the letters surrounded by no fewer than 10,600 red dots, which have been included simply to produce a soothing background to the characters. It has been calculated that putting in the dots alone must have taken, at the very least, six intensive man-hours. This artistic device was originally a characteristic of Coptic manuscript illumination, whence it passed into the Byzantine tradition by the sixth century, before becoming incorporated further in the techniques of the Celts.

It is a matter of astonishment that some of these artworks are with us still. At the dissolution of the monasteries in the sixteenth century, the *Book of Durrow* passed into the hands of a MacGeoghan, who lived nearby in County Offaly, and got into the habit of doctoring his sick cattle by plunging the relic into water, which he then allowed the animals to drink, a treatment which eventually resulted in many pages being holed by damp rot. The *Book of Kells,* the most celebrated of all such texts ('the most sumptuous of the books to have survived from Europe's early Middle Ages', in one expert opinion), had an even more perilous passage to the twentieth century. There is a school of thought which believes that it was, in fact, created in the monastery on Iona sometime in the middle of the eighth century, and that it was taken to Kells in County Meath to protect it from the Vikings, who were raiding the Scottish islands and west coast a hundred years later. This is certainly plausible, given that Kells was an Irish daughter house of Iona's and did, indeed, take custody of reliquaries associated with Columba, when these were transferred for safety's sake in 877. The book's sojourn in Kells, however, whether or not it originated there, certainly did not guarantee its preservation, for that area was also invaded in the tenth century. The volume survived, only to be stolen from the church in 1007, turning up in a ditch several weeks later, minus the shrine of precious metal and jewels in which it was kept.

The final hazard came at the hands of an eighteenth-century book-binder, not the first craftsman who had been asked to take the volume into his care. But this one trimmed its leaves disastrously, to the extent that some margins lost nearly an inch, and illustrated pages were damaged in the process. Worse, it is thought that there must have been 370 folios in the book when it was created by the monks, whereas today only 340 remain. At least we have those, and a glory of other books like *Durrow* and *Lindisfarne* to tell us of what once was. It is obvious that, through warfare, pillage and other tragedies, we have lost many more of these illuminated manuscripts than we shall ever know. But that any of them have survived to our own time is nothing less than a miracle in western civilisation.

8

The Scoti

Reference to anyone as a Scotus (or Scottus), to a group of such people as Scoti (or Scotti) at this period can be confusing, in view of subsequent topographical history. These were, in fact, popular expressions by others for what we now call the Irish; and the Scots of the historical Scotland were not to be so named until early in the twelfth century. About 500, the future kingdom of Scotland was occupied by three separate racial groups. The warlike tribe of Dál Riata had lately invaded from their stronghold of Dunseverick in the northeast of Ireland (a land which in its entirety was identified by the Romans as Hibernia, by its own people as Eriu, or Hériu), had brought the Gaelic speech with them, and had occupied modern Argyll and southern Inverness, which included Kintyre, Cowal, Lorn, together with islands off their shores; an area which soon went under the name of the invaders (gradually corrupted to Dalriada). To the south and east, Britons occupied Strathclyde, between the Cheviot Hills and a boundary formed by the river Clyde and the Firth of Forth; and everywhere above this natural frontier, outside Dalriada, belonged to the Picts. The aggressive

Scotti steadily pushed their boundaries beyond Dalriada and, as a result, the whole area above the Cheviots eventually became Scotland.

The confusion that sometimes arises from these comings and goings (not least among continental Europeans, who have often been misled into believing that Irish monks wandering the medieval continent were, in fact, Scotsmen) is illuminated in the names of two individuals. John Scotus Erigena was an Irish philosopher who may have been driven from his native land by Viking invasions, and was made welcome at the court of Charles the Bald, Charlemagne's grandson. He engaged in some notable disputes on the nature of God, is considered a neo-Platonist of considerable originality but scarcely any influence, though Peter Abelard is supposed to have been affected by his teaching. There is a legend that Erigena died (*c.* 877) at Malmesbury, where his pupils are supposed to have stabbed him to death with their pens; but there is no acceptable evidence for this. Some four hundred years later Duns Scotus appeared, another philosopher and a Franciscan friar, the first figure of any weight to defend the dogma of the Immaculate Conception against the scepticism of certain recognised theologians. It was symptomatic of subsequent (sixteenth-century) humanist and reforming contempt for the position he and his sympathisers adopted, that the word 'dunce' was coined in mockery of him. But he (*c.* 1264–1308) was a genuine Scot, as we understand the term today, having been born in the Border country of Roxburghshire.

8

Antony and Cassian

The two texts which excited Fionán when he discovered them in the Clonfert scriptorium were Athanasius's *Life of Antony* and the *Conferences* of John Cassian. The first of these writers was a fourth-century bishop of Alexandria who is remembered best for his long opposition to the Arian heresy (it denied the essential divinity of Jesus Christ, making Him subordinate to God the Father in the Trinity), as a result of which he was twice forced into exile. On the first occasion the Emperor Constantine dismissed him to Gaul, when Athanasius overplayed his hand by threatening a dock strike which would have cut off vital corn supplies to Constantinople, unless he received imperial support in the controversy. On Constantine's death, he returned to Alexandria but subsequently fell foul of the emperor's son Constantius, who banished him again, this time to the Egyptian desert; which is where he encountered Antony, and fell under his spell. Athanasius was long associated with the credal statement that bears his name ('Whosoever will be saved: before all things it is necessary that he hold the Catholick Faith . . .' as it begins in the Anglican Book of Common Prayer); but since the seventeenth century this has been doubted, and the Athanasian Creed is now thought to have been the work of his contemporary, St Ambrose of Milan.

John Cassian (c. 365–435) came about half a century after Athanasius and was a monk of Scythian origin, possibly born in one of the Latin-speaking areas of what is now Romania. He entered a monastery in Bethlehem as a young man, before spending several years in the Egyptian desert, where he absorbed everything the Fathers there had to offer him, without swallowing it all uncritically. He then went to Constantinople whence, because he took sides in various theological disputes and fell out with the dominant parties, he later moved to Rome, then to Provence: this was the way of the world in the early centuries of Christianity. In Marseilles he established a religious community for men and another for women, modelled on the versions he had studied in North Africa; and he left his mark on any number of hermitages that were beginning to spring up along that Mediterranean coast. There can be little doubt that he greatly influenced the monastery lately founded by Honoratus on the island of Lérins, where Patrick was formed and made deacon before returning to Ireland. It was Cassian who regarded the Psalter as the most important textual aid for a monk—or a community of monks—at prayer. He would be the source whose authority caused sixth-century Irish monks to learn the 150 psalms by heart.

In Gaul, Cassian began to write. In his *Institutes* he concentrated on the organisation of a religious community, producing a model that Benedict would elaborate upon a little later in his famous *Rule*. In his *Conferences,* Cassian was much more concerned with the spirituality of the life, in twenty-four discourses which he put into the mouths of fifteen Egyptian Fathers. They treat of such matters as telling the absolute truth, and whether it should take precedence over other forms of charity; the degree of friendship that should be allowed, even cultivated between monks; causes of quarrel and methods of handling anger; ways of channelling sexuality for the enrichment of the celibate life; and so on. At the heart of Cassian's spirituality were Conferences

Nine and Ten on the subject of prayer. Its aim, he said, was the constant and undistracted direction of the mind towards God, and purity of heart was its necessary precondition. The greatest obstacles to prayer were, in this order, sin, anxiety, pride, an absence of tranquillity. He pointed out that every individual prays differently, but that all versions of prayer will (or should) contain resolution, penitence, intercession for others, gratitude. One other topic exercised Cassian as much as prayer. Was it, he asked himself, possible to achieve perfection in the religious life? He doubted it, which is why he so antagonised followers of Augustine of Hippo, who propounded the doctrine for all monks and nuns, and conveyed the feeling that any failure to achieve perfection was something close to betrayal of their vocation.

Cassian, in short, was a moderate man who discouraged all exaggerated forms of religious experience, including rumour of supernatural occurrence. When a popular biography by Sulpicius Severus ascribed to Martin of Tours the ability to work the most prodigious miracles, the intention of its author probably being to demonstrate that Gaul's holy men were superior to Egypt's, Cassian deplored it and its effect on a credulous people. As a moderate, his influence was to be most marked in its effect upon Benedict, whose *Rule* also reacted against forms of extreme behaviour. Benedictine houses had readings of the *Conferences* (collationes) before the last office of the day, eventually during a light meal which preceded compline. It is from this regime that the English word 'collation' is derived, as well as the Italian 'colazione'.

8

The Desert Fathers

The Desert Fathers who inspired both Athanasius and Cassian, and through them became the model for early Irish monasticism, were men who had sought the wilderness for a variety of reasons. The desert had an intrinsic significance for anyone within the Judaeo-Christian tradition, going back to Abraham, and there can be little doubt that some had simply responded to this deeply biblical pull. Others probably saw it as a way of avoiding taxation or military service in the imperial army of Rome. Many were certainly in flight from religious persecution, a recurring feature of the Empire until Constantine championed Christianity in 313. Typical of these pogroms was the one ordered by the Emperor Decius, who reigned for only two years in the middle of the third century that, for Christians, were calamitous. His first act was to execute Fabian, Bishop of Rome, in 250; six months later, he required all citizens of the Empire to furnish proof that they had lately made sacrifice to his divine person, and killed many thousands who could not or would not produce certificates. (In much the same frame of mind, Henry VIII of England would later demand oaths of unquestioning obedience from his subjects, when many of these became too obviously agitated at his intention to shed his first wife in favour of Anne Boleyn; Sir Thomas More and Bishop John Fisher of Rochester being only two of the many who were put to death because they could not in all conscience so oblige the King's Commissioners.) One fugitive from the persecutions launched by Decius was Paul of Thebes, who lived in a cave near the Red Sea until his death at the age of 113, *c.* 340, and who is generally accepted as the first Christian hermit. The primary impetus for those Christians like him, who remained in the desert when all threats had receded, was clearly a belief that in the huge silence of the great emptiness they could best respond to the claims laid upon them by the Gospel.

Colonies began to form in various parts of the Egyptian interior. At Nitria, a long valley some sixty miles south of Alexandria, there were reckoned to be five thousand zealots by the end of the fourth century, each living according to personal inclination. Two or more anchorites often dwelt together in cells that were no great distance apart. They generally worked from dawn until mid-afternoon, when they returned to their dwellings to sing psalms and hymns; but on Saturday and Sunday, all gathered at a church in the middle of the valley for common worship under the leadership of priests. Nearby were three palm trees, each with a whip hanging from it: one for transgressors in the community; one for robbers who might be caught there; one for any itinerant who might merit—or seek—flagellation. A guesthouse stood close by, where a visitor could stay as long as he chose, though after a week of idleness he was expected to work in the garden, in the bakery or in the kitchen. Labour was regarded as the great antidote to accidie, the weariness of heart that has periodically afflicted every religious throughout history, making him or her dislike the very idea of community, cell, individual brothers or sisters, vocation itself.

Only a few miles from Nitria was Cellia, where six hundred hermits dwelt in huts so constructed that no inhabitant could either see or hear any neighbour: the huts usually had no more than one room, and each hermit made his own out of sun-dried bricks in a single day.

Apart from communal worship at the weekend, these men had nothing at all to do with one another. Among them was Macarius the Egyptian, who lived at Cellia periodically, and who became notorious for adopting any eremitical practice he heard of during his excursions elsewhere and then surpassing it when he returned to his hut: as when for seven years, he ate nothing that had been cooked. Another hermit with the same competitive urge put his daily bread in a narrow-necked jug and consumed only what he could extract in one grasp. All hermits tended to eat nothing but dried bread and green herbs, each liberally salted, to preserve them and to make them more edible. Trying to go without sleep, or sleeping outside in the bitterly cold desert nights, were commonplace mortifications of the flesh. An old man named Didymus used to tread on scorpions with his bare feet.

A day and a night's journey across the desert to the southwest of Nitria was Scete, founded in 330 by another Macarius (the Alexandrian), who lived there for the next sixty years. Because he had the gift of healing he was constantly plagued with visitors, and to escape them he excavated a tunnel from his cell to a cave half a mile away, leaving one of his two closest disciples to cope with sick people seeking a cure. Another celebrated member of the community was Paphnutius, a priest who became renowned for his punctuality at the weekend worship, though he had to walk five miles to get to church from his hut. Then there was Moses, an Ethiopian, who had been both a government official and a robber before becoming Christian. He was constantly troubled by nocturnal temptations and to combat them, it was said, he remained standing in his cell at night for six years, without ever closing his eyes. When even this failed to subdue his sexuality, he took to going round all the cells in the colony at night, surreptitiously taking their empty water jugs, and refilling them from a source two miles away.

One of the most celebrated colonies was founded by Pachomius, who was born into a pagan family in Upper Egypt, and pressed into Constantine's army. He and other soldiers were befriended by Christians who gave them food and drink, which caused Pachomius to ask for his discharge and seek baptism. He joined a group of anchorites but, *c.* 320, detached himself to live alone for a while in Tabennisi, a deserted village beside the Nile, two hundred miles upstream of Lycopolis (the modern Asyut). And there, under his leadership, a novel form of community was established. All his predecessors had allowed their followers to work out their own routines, without being pressed into any conformity. By deciding that there should be an inflexible rule of behaviour for all at Tabennisi, Pachomius is quite properly seen as the begetter of the western monastic tradition, with the common coenobitical life at its centre, in contrast to the eremitical life of the individual anchorite. His *Rule,* written in Coptic to start with, was translated into Latin by Jerome in 404 and thereafter had immense influence, upon Cassian among others. Cassian's *Institutes,* which in turn affected Benedict's approach to monasticism, are clearly a refinement of Pachomius.

Pachomius soon had a hundred monks with him, and by the time he died, he had under his obedience no fewer than nine monasteries for men and two for women. Each monastery was surrounded by a wall, which might have any number of structures inside it, quite apart from the church. One establishment contained between thirty and forty buildings, each housing between twenty-two and forty monks, with an individual cell for each, a common refectory, a guest-house and a vegetable garden. In charge of every monastery was an Archimandrite (known as pater or abbas to his brethren) who nominated his successor before he died. Novices of all ages and condition were attracted, including slaves, though these were admitted only with their master's consent; and priests were unwelcome unless they submitted absolutely to the monastic authority. Every monk dressed alike in a sleeveless linen tunic which came down below the knees, and was secured with a girdle of linen or leather. A sheepskin or goatskin cloak went over that and, above this, a hooded cloak with a sign on the back, indicating which monastery the monk belonged to. He went barefoot at all times except when travelling, in which case he would wear sandals and carry a staff. In the monastery there were common prayers at dawn, midday, sunset and midnight, Saturday and Sunday mass with communion, and spiritual instruction by the superior to the whole community twice a week. There were two meals a day, at noon and in the evening, but both meat and wine were strictly forbidden. Work was highly organised, because Pachomius believed not only that it was necessary to sustain the community; he also thought it a very healthy ascetical exercise. Apart from communal chores around the buildings and in the garden, therefore, his monks generally employed themselves weaving mats.

Antony's was the name that rang above all others. He was born into a prosperous Christian family near Heracleopolis, on the Lower Nile, in 251, a shy child who never learned to read his native Coptic, or Latin, or Greek. When his parents died, he sold the property, gave the proceeds to the poor, put his sister into a house of virgins, and began to live as an ascetic nearby. His days were spent in prayer and work, earning him enough to live on, with a little left over to be distributed as alms. Because he was illiterate, he memorised Scripture as it was read out in church, and he began to visit distinguished recluses in order to take note of their virtues. 'He observed the graciousness of one,' says Athanasius,

the eagerness for prayers in another, he took careful note of one's freedom from anger, and the human concern of another. And he paid attention to one while he lived a watchful life, or one who pursued studies, as also he admired one for patience, and another for fasting and sleeping on the ground. The gentleness of one and the long-suffering of yet another he watched closely. He marked likewise the piety towards Christ and the mutual love of them all. And having been filled in this manner, he returned to his own place of discipline, from that time gathering the attributes of each in himself, and striving to manifest in himself what was best from all.

He ate only once a day, after sundown, no more than salted bread and water; and he slept on a mat or on the ground, often spending the night

in vigil. After fifteen years like this, he decided he needed greater solitude and withdrew in 285 to the other side of the Nile where, at Pispir
on the eastern bank (the modern Dar al-Maymun), he found the old
fort which made Fionán smile when he read that it was crawling with
reptiles, such as Patrick had despatched. For twenty more years he
lived there in complete isolation, his bread brought to him by friends
every six months while he prayed and wove mats. This proved irre-
sistibly attractive to many, who forced themselves upon him as disciples, and built cells close by, where they imitated Antony's way of life,
and met from time to time to hear an address from him on how they
should conduct themselves. The Emperor Maximin was putting down
the Christians again, though Pispir was too remote to be affected; so
Antony with some of his disciples went to Alexandria in the hope of
martyrdom, which eluded them in a manner that Athanasius describes
with shining admiration.

He yearned to suffer martyrdom, but because he did not wish to hand himself over, he rendered service to the confessors both in the mines and in the
prisons. In the law court he showed great enthusiasm, stirring to readiness
those who were called forth as contestants, and receiving them as they un-
derwent martyrdom and remaining in their company until they were per-
fected. When the judge saw the fearlessness of Antony and of those with
him, he issued the order that none of the monks was to appear in the law
court, nor were they to stay in the city at all. All the others thought it wise to
go into hiding that day, but Antony took this so seriously as to wash his
upper garment and to stand the next day in a prominent place in front, and
to be clearly visible to the prefect. When, while all marvelled at this, the prefect, passing by with his escort, saw him, he stood there calmly, demonstrating the purposefulness that belongs to us Christians. For, as I said before, he
also prayed for martyrdom. He seemed, therefore, like one who grieved because he had not been martyred, but the Lord was protecting him to benefit us and others, so that he might be a teacher to many in the discipline that
he had learned from the Scriptures. For simply by seeing his conduct, many

aspired to become imitators of his way of life . . . When finally the persecution ended, and Peter the blessed bishop had made his witness, Antony departed and withdrew again to the cell, and was there being daily martyred
by his conscience, and doing battle in the contests of the faith.

Disciples now began to arrive in such quantity that Antony fled into
the desert, joined a Bedouin caravan and travelled with them to a
place in the hills not far from the Red Sea. And there he would remain
until he died in 356 at the age of 105, occasionally returning to Pispir
to give the community advice, once venturing as far as Alexandria
again, in order to support Athanasius in the Arian controversy. At his
death, he left his biographer a sheepskin tunic and the cloak that he
had slept on at night for many years, having given instructions that he
was to be buried in a secret place, so that he should not become the object of a cult. This was a vain hope, for no reputation among the holy
men of Egypt carried as far as Antony's.

Yet collectively, the Desert Fathers had an extraordinary impact
upon the Christian world in their own age and the period immediately following it. Cassian and Athanasius were not the only ones who
succumbed to the appeal of the men and women who had chosen to
isolate themselves in this rigorously self-denying way. Another chron-
icler was Tyrannius Rufinus, a presbyter from northern Italy and an
important translator of Greek theological works into Latin who, towards the end of the fourth century, journeyed through Egypt with
six companions in order to inspect the anchor-holds and the monasteries of such fathers as were prepared to receive them. Their findings
were written up by Rufinus as *Historia Monachorum in Aegypto* and
became the standard text on the topic, to sit alongside works by Cassian and Athanasius. All three authors were addressing themselves to
the developing religious communities of western Christendom, to the
men and women who thereafter looked to the Desert Fathers for example. No country was to be more affected by this than Ireland. The
association of the desert with the deeply devoted religious life became

implanted in the Irish psyche to a remarkable degree. Individuals thereafter were consumed with a desire to find their own desert in the soggy landscapes and along the rocky coasts of the eastern Atlantic seaboard. A lasting testimony to this urge are the Irish place names which today incorporate the word or a variant—Dysart (Westmeath), Desertmartin (Londonderry), Disert Oenghusa (Limerick), Killa-dysert and Dysert O'Dea (Clare), and others.

10

St Antony in Art

The most unpredictable effect of Antony was upon the history of western art; or, precisely, it was the effect of the torments that Antony endured during his initiation into the anchorite's life. Artists became captivated by the subject from the fifteenth century at least, and have never left it alone for very long since, up to and including the twentieth century. From Matthias Grünewald and Albrecht Dürer, to Max Ernst and Odilon Redon, by way of Tiepolo, Teniers and many, many more, generation after generation of painters have sometimes seemed to regard this reclusive figure and his legend as a test piece of their art. It has been argued that this followed from the saint's great popularity after the foundation in 1095 of the Hospital Brothers of St Antony, an order which dedicated itself to the treatment of contagious diseases, one of which was known as St Antony's Fire (ergotism, which produced intolerable pain, disfiguration of the skin and some-times gangrene). But that argument, surely, would hold good only for the Middle Ages.

One of the first attempts to picture The Temptation of St Antony the Abbot may or may not have been made by Fra Angelico: experts have never been able to agree whether the work was executed by the great friar himself or by a pupil working under his direction. Whoever the painter was, he produced what is possibly the most uncommunicative treatment of this subject that was ever made in the graphic arts. The saint (dressed, interestingly, as a Benedictine monk rather than like someone from Angelico's own Order of Preachers) is certainly in no wilderness, for the surrounding hills are covered with verdant trees, and there are several buildings in the background, including a large and castellated shape with numerous towers. A few animals are included, but they appear to be wholly supine, and it is a little difficult to see exactly what Antony is being tempted by or with, especially as the expression on his face suggests blankness rather than turmoil, and the right hand looks as if it is raised in casual salute rather than in supplication for divine assistance, or as a gesture to ward off the Devil.

Hieronymus Bosch, predictably, was much more lurid than this when he turned to Antony, which he did more than once, though never as spectacularly as in the great triptych, with its brilliant russets and reds, its subtle greys and blues, which is now housed in Lisbon's Museu de Arte Antiga. He had worked up to this masterpiece with a drawing of *Studies for the Temptation,* and with another triptych, *Hermit Saints,* where Antony appears on one panel, followed by Jerome and Giles on the others. In the Lisbon work, the left panel shows Antony being helped to safety after being beaten senseless by demons, and one of his companions is supposed to be a self-portrait of Bosch. In the right-hand picture, he is being tempted by a naked Devil-Queen (who is, however, coyly covering one breast and her pubic area, quite unnecessarily when the saint's eyes are painstakingly averted) while other seductive figures are offering him all the delights of gluttony. The centrepiece is simply a comprehensive review of every vice and evil that was employed by the Devil to divert Antony from prayer and the path to holiness, while the saint kneels in the midst of this bedlam before a tomb in which Christ crucified is half-hidden, beside a figure whose hand is raised in benediction.

It is, of course, the immense detail as well as the artist's powers of invention and extravagant imagination that makes the fantasy so compulsively readable: the monsters that harry Antony even as he is borne to safety in the sky, the tonsured monk reading Scripture with spectacles balanced on his swinish snout, the man whose kneeling body forms the roof and entrance to a brothel, the seashore littered with corpses from a shipwreck, the dragon swimming in the moat of the Devil-Queen's castle, people offering gifts to an enthroned ape, others swilling liquor with their backs to Christ. There are recognisable humans, carefully drawn fishes, birds and mammals; and there are creatures that are neither one thing nor the other but clearly diabolical—a bird with a long beak that ends in a smoking trumpet, the infant in the arms of a woman who terminates in a lizard's tail and who is sitting on a giant rat, the half-donkey completed as a jug, the half-galley half-swan sailing across the sky, the plucked goose with a sheep's muzzle and clogs on its feet . . . one could go on and on. Bosch's version of the Temptations is infinitely more harrowing than Athanasius's written description of them, because it so graphically represents all the fearful images that tormented the medieval mind and inclined it to obedience before the altar and in the presence of priests. When set alongside the sceptical Dutch genius, Athanasius can be seen as no more than a tone-deaf votary.

Not even Pieter Brueghel seriously competed with Bosch in his own treatment of the Temptation, a drawing which is thought to be his earliest surviving composition. This is another work that has baffled experts, no one having come up with an agreed interpretation of the dominant feature, a head lying in water, with all manner of activity issuing from various orifices, most of them disagreeable and some intimidating. This is seen by many as a caricature of the corrupt Church, on which the diminutive figure of Antony has conspicuously turned his back, bottom right of the picture, where he is kneeling dolefully, supported by his stick. A cameo in the foreground—man on a barrel, jousting with a mop—prefigures an almost identical detail in one of Brueghel's major paintings, *The Battle Between Carnival and Lent*.

Other things in the composition are also familiar, like the gaping fish balanced atop the detached head, the flock of birds in the background, the human figures gymnastically up trees or doing mundane things with jugs, trumpets, fishing rods; and the most interesting thing about the drawing as a whole is not its place in the iconography of St Antony, but its intimation of what was to come later in the development of a great artist.

Michelangelo, too, is supposed to have produced a Temptation after seeing an engraving of the subject by the German Martin Schongauer, in which demons are ravaging the saint with their hands; but as this is one of the master's works that have been lost, we only have Vasari's word for it that Michelangelo 'copied it [the engraving] with a pen in a manner which had never before been seen. He painted it in colour also . . .' Other Italian works have survived. The sixteenth-century Bolognese Annibale Carracci varied the subject by painting *Christ Appearing to St Antony Abbot during his Temptation,* after reading not Athanasius, but a hagiography of several saints entitled *The Golden Legend,* which was composed by the thirteenth-century Dominican Archbishop of Genoa, Jacob of Voragine. Clothed in blue and pink, Carracci's Christ is airborne, supported by three putti while, below, Antony is recumbent in the lee of a rock with an open book on his lap, a lion snarling at his feet, a devilish humoid with a bat's wings about to claw him, other diabolic figures in the shadows, one brandishing a snake; and all except the lion looking upward to the heavens. The chiropterous creature, with claws hideously extended at the end of limbs that were all bone and membrane, was to crop up repeatedly after Schongauer first introduced this spectre, when artists wished to threaten Antony with something that made everyone shudder. It was the central feature of the Temptation by the seventeenth-century Neapolitan Salvator Rosa, a nightmare combining bird and reptile, with its extremities splayed over the saint as if about to mount him, so appalling that Antony shrinks against the ground and holds out a cross to stave off the impending obscenity.

The most rewarding nexus of Antonine art and literature came when Gustave Flaubert was struck by a version of the Temptation attributed to the younger Pieter Brueghel, which he came across during his sister's honeymoon (!) in Genoa; and then acquired another treatment of the subject by the seventeenth-century etcher Jacques Callot, which pictured the saint attacked by monsters breathing flames, while other creatures danced around him in a frenzy. Flaubert hung a copy of the Callot on his wall in 1846 so that he could study it the better, and it seems that from this fixation, as well as from his own torment, he gradually composed over the next twenty-seven years (though extracts were published from a first draft in 1856) *The Temptation of St Antony,* the novel that he himself regarded as 'l'oeuvre de toute ma vie'. This did not much resemble Athanasius's portrait of their subject, which depicted him as gentle, humble, patient, unsophisticated, resolute; whereas Flaubert's Antony is cantankerous and bitter, greedy and envious. That may well be unfair to the historic Antony; but the novelist unquestionably illuminates an area that the fourth-century apologist was simply not equipped or inclined to enter. Flaubert's Temptation is a brilliant study of religious dementia, a revelation of someone becoming unhinged by extreme sanctity.

And it worked, artistically and as an inspiration to others it worked, because a number of painters who read the novel then reached for their brushes and tackled the theme themselves. The French Romantic Henri Fantin-Latour was one, and the Belgian James Ensor was another, though he was almost certainly stirred also by what Bosch and Brueghel had done, for he was a devoted student of their work. Most notably, Cézanne was impelled by Flaubert to paint the Temptation yet again, having already been attracted to the subject before, including himself in one composition as a marginal figure looking on anxiously, his Antony in another being modelled on his friend Emile Zola.

Like Tintoretto, Patenier and Mabuse in the sixteenth century, like Millet and Fantin-Latour just before in the nineteenth, Cézanne saw Antony's temptation as carnal and nothing more. So did Stanley

Spencer in 1945. The saint lies in an opened table-tomb (presumably in Cookham churchyard) surrounded by a dozen or more naked women, all with big breasts and wide, fleshy hips, the basis of Spencer's fantasies since his youth (two or three of them are strikingly like Hilda Carline or Patricia Preece), and all deeply interested in rousing Antony from his depths. Twelve months later, Salvador Dali produced the most arresting interpretation of the legend since Hieronymus Bosch, though he had clearly been stimulated by Salvator Rosa more than by any other predecessor: his Antony is caught in an almost identical pose (but naked), holding out the cross in the same fashion to ward off a rearing brute (an unmistakeable horse in this case); and there the comparison ends. A number of other stilt-legged creatures also inhabit the grey and almost featureless flat desert, but these are all elephants, bearing various pieces of masonry: a couple of phallic columns, a basilica with female torso exposed in its doorway, a fountain crowned with a buxom woman fondling her own breasts. Dark clouds are looming and within them is a building that could equally be penal or religious in its purposes; while distantly on the desert floor, a clothed monk is also brandishing a cross at the spectral figure of a man, who looks as if he might be beseeching rather than posing a threat.

And that touches on the most notable thing about the artistic treatment of Antony, from first to last. Though every single composition, from Fra Angelico onwards, refers to the Temptation(s) of its subject, the fact is that with the sole exception of Hieronymus Bosch, Antony does not appear to be tempted by anything except women and the possibilities of sex. Bosch alone hints at several forms of seduction by depicting avarice, gluttony and other falls from Christian perfection. What the saint experiences most of all, however, except at the hands of the painters who concentrate on lust alone, is the threat of something unpleasant, and sometimes terrifying, happening to him. As often as not, the paintings were doubtless critiques of the Church, for which the legend of Antony provided a dramatic excuse.

11

The Irish Kings and Tara

The structure of Irish society, at the time when Fionán and his companions were setting forth from Brendan's foundation at Clonfert, was essentially tribal, with a recognised hierarchy of kingdoms of varying size: there may have been as many as a hundred tuatha all told. Lowliest of the chieftains was the rí túaithe (petty king), and there were other levels of power, including the rí coícid (king of a province), before the summit was reached in the person of the ard rí (the high king), who ruled from Tara in County Meath, and was the supreme chieftain with authority throughout the land. His identity depended almost entirely on which dynasty had the upper hand in and among the five principal kingdoms of Ulster, Connacht, Leinster, Meath (which was eventually absorbed by Leinster) and Munster.

By far the strongest, more often than not, were the Ui Niall (or Ui Néill, which became eventually O'Neill), who could look back to the Solomon of Irish royalty, the third-century Cormac mac Airt, for their ancestry. There were two branches of the clan, the southern Ui Niall in Meath and the surrounding midlands, and the northern Ui Niall in Ulster. In the seventh century, no fewer than eight members of the southern Ui Niall were kings in Tara, until they were superseded by their rival kinsmen, the Clann Cholmáin, in 743; which, apart from an interlude from 944 to 956, would exclude them from Tara forever. From the 740s onwards, the ard rí alternated from either the Clann Cholmáin or from the Cenél Eogain, who were another clan of the northern Ui Niall.

Leinster was first ruled by the Ui Dúnlainge, who were lodged in the vale of the Liffey and the northern plains of their kingdom. They would maintain strong links with Brigid's dual monastery at Kildare, whose successive abbots and abbesses tended to share the royal blood, until the Ui Niall made war on the province in 780 and thrice more in the next forty years. Thereafter, it was the Ui Niall who appointed the kings of Leinster.

In Munster the tribe of Eogonachta held sway, divided into two rival groups like the Ui Niall, one dominant about Killarney and the south of the kingdom, the other round Cashel, Glenworth, Knockaney. Both branches, however, prized their Christian antecedents, claiming that angels had revealed their royal stronghold at Cashel, proud that a forebear had been baptised by Patrick himself. Some became priest-kings, others hereditary abbots, especially close to the important monastery of Emly. A ninth-century commentator remarks on the gentleness of their rule compared with that of the Ui Niall; but this was to be their undoing. Pressed for concessions by the Ui Niall, harried further when the Vikings appeared, divided among themselves, the Eogonachta crumbled in the tenth century, giving way to the Dál Clas, who raided from the region of the lower Shannon and swiftly built up a southern power base from which, one day, they went forth to capture Viking Limerick and became the first Irish urban dynasty.

Western kinsmen of the Ui Niall were the Ui Briúin of Connacht, whose rise began in the seventh century and who dominated from 725. Realising the growing power of the Church at that time, and especially the influence of the great monasteries, they were clever enough to patronise Clonmacnois, which became one of their dynastic churches; at the same time, banking on an eventual ascendancy of the bishops, they maintained excellent relations with Armagh by imposing its ecclesiastical tax in their own lands. And the strategy paid off, though economically Connacht would always be the poorest province in the country. The Ui Briúin achieved a stable line which presently, in the twelfth century, brought them the kingship of all Ireland, which was still associated with Tara, though no more than nominally by then.

H. V. Morton once memorably wrote of Tara that 'Ireland is full of old unhappy things that strangely shake the heart; and this mound of earth is one of them, lonely, remote and withdrawn like "something left on earth after judgement day."' That was in 1930, and the seat of Ireland's ancient royalty has changed not at all since then; apart from what Peter Harbison, a considerable authority, has described as 'the atrocious statue of Patrick'. Mitred on his plinth, crozier firmly grasped in one hand, the saint now watches over the haunted earthworks, though with the advent of Christianity, Tara's days were numbered as an active headquarters of the ard rí, and Mael Shechlainn abandoned it to mere symbolism in 1022. Bardic tradition insisted that it was built 'in the year of the world 3922' by Ollamh Fodhla, its first ruler, and that from the time of the Fir Bolg until Patrick's day, no fewer than 136 priest-kings in succession were crowned and subsequently dwelt in this place, which was hallowed by religious as well as regal significance. Nothing but earthworks remain of the banquet hall, the royal enclosure, Cormac's house or the hall of assembly, where the synods arrived through fourteen entrances from every corner of the land, along the five great highways that all converged on this spot: down from the northeast through Dundalk, up from Ossory and eastern Munster, by the Slope of the Chariots from the northwest, from distant Connacht in the west and, shortest journey of all, along the Wicklow road. But originally there must have been structures of wood or wattle and daub in Tara, at least from AD 218 to 260, when Cormac ruled even-handedly, giving equal attention to the encouragement of poets, bards and chroniclers, and what was conceived as a sacred duty of the Irish to plunder at every opportunity the islands lying just across the water (one of his expeditions among the Britons is reputed to have lasted three full years).

Below the level of the tribal chiefs, the hierarchy was a little more flexible, insofar as the acquisition of wealth enabled a family to rise in Irish society, though this might take three generations; whereas breeding was the conclusive requirement for kingship. There were nobles and there were commoners, and the distinction between them, apart from those of birth and wealth, usually depended on how many clients each could muster, bound to them by contract of some sort, owing services—ploughing, harvesting, building—in return for a form of tenancy and for protection against violence, or in any other kind of dispute. The commoners were usually freemen who occupied land which, in the case of a bóaire (literally cow-lord), would cost him a cow and other payments each year. Quite apart from these outgoings, however, the commoner was expected to entertain the lord when he embarked on a winter tour of his lands and, with his retinue, feasted at the expense of each client in turn, thus saving a great deal from his own budget. The client relationship also meant that the lord could rely on a number of men to assist him in battle, and these were rewarded with a share of the spoils. Below the commoners came the landless men who were bound to the soil; and below them were slaves, who had been taken (as Patrick was taken) on raids abroad, or in inter-tribal warfare.

Women were, of course, merely incidental to these arrangements: they were there for their breeding capacity and to drudge around the home. There were two forms of marriage, one sanctified by the Church, the other made more traditionally with inherited bride-wealth. But divorce and remarriage in the traditional mode were common, and the Irish lordlings often had one wife after another, which assured them of numerous heirs and accomplices.

12

Three Martyrdoms

The martyrdom that Fionán and Brendan had in mind clearly involved the death of the martyred person; but this was only one of three different forms that the early Irish Church recognised. It has been convincingly argued that because of the exceptional circumstances in which Christianity came to Ireland—because, that is, no blood was shed at its coming, nor any at all until the Vikings arrived at the end of the eighth century, and began slaughtering monks and other Christians without compunction—the Irish constructed for themselves other forms of suffering for the faith that they could think of as martyrdoms. Martyrdom had been inseparable from Christianity until it reached Ireland; the faith was actually founded on the biggest martyrdom of all, and adherents doubtless had a deep psychological need to feel threatened with extreme punishment, to reassure themselves not only that they shared the attested pedigree of the early saints and martyrs, but that they were worthy followers of the martyred Christ.

A directive written in Irish, probably before the end of the seventh century, represents the position that the early Church there had arrived at by then.

Now there are three kinds of martyrdom, which are accounted as a cross to man; to wit, white martyrdom, green martyrdom and red martyrdom. White martyrdom consists in a man's abandoning everything he loves for God's sake, though he suffer fasting or labour thereat. Green martyrdom consists in this, that by means of fasting and labour he frees himself from his evil desires, or suffers toil in penance and repentance. Red martyrdom consists in the endurance of the cross or death for Christ's sake, as happened to the Apostles in the persecution of the wicked, and in teaching the law of God.

The concept of degree, in fact, seems to have originated in Athanasius's reference to Antony being 'martyred by his conscience' when he failed to find the death he had sought in Alexandria during the Maximinian persecutions. Sulpicius Severus added his weight to the notion by insisting that although St Martin of Tours was never put to death for Christ's sake, he endured so much else only just short of that, in hunger, insults, sleeplessness, nakedness and other disagreeable substitutes, usually of his own choosing, that the result amounted to a form of martyrdom. It is possible that the Irish drew directly on both these sources for their own degrees of martyrdom, though it is thought more likely that the idea came to them in the first place (along with much else) from the island of Lérins, where Patrick was formed, and with which a lasting connection was made; it has been well described as 'the holy isle par excellence of Europe' by the fifth century.

Another saint whose introduction to the ascetic life occurred at Lérins was Caesarius of Arles, who commonly preached about the possibility of living martyrdom, almost a century after Patrick. 'For it may be said without disrespect for those who lay down their lives during the persecutions, that to afflict the flesh, overcome lust, resist avarice and triumph over the world is to go far on the way to martyrdom.' It has been suggested that the route taken to Ireland by such a pronouncement would be in the memory or even in a manuscript borne by some monk travelling across Gaul and by sea from Britanny to Wales; most likely to the monastery of Ynys Pyr (Caldey), where St Cadoc trained, thence to

Nant Carfan, which he later founded, and from there to Clonard, whose founder Finnian sat as a novice at Cadoc's feet.

The idea of modified martyrdoms may not have originated with the Irish, but it was they who introduced the terms 'red', 'white' and 'green' to distinguish the three forms; and no monks of any other land took up the concept as devotedly as they did. A Litany in the *Book of Leinster* refers to some seventh-century monks as martyrs, when they simply lived under a particularly rigorous abbot; and when Columbanus wrote the *Rule* that so discomfited some of his continental brethren, he applauded 'the felicity of martyrdom', making it clear that he was referring to something in the monastic life that fell short of being brutally put to death.

The most extreme form of this was, of course, the life of the hermit, modelled on the example set by Paul of Thebes, whose only garment was made of palm leaves, and who ate nothing for most of his life but the half-loaf of bread that a raven brought him daily for seventy years; except on the occasion when Paul was visited by Antony, when the bird thoughtfully supplied a full loaf. The Irish seem to have concluded that such a level of mortification should only be attempted after long preparation for it in a community of ascetics; and so they sometimes effected a compromise between the hermitage and the monastery by establishing the small semi-eremitical community, which is what Brendan urged upon Fionán as a matter of prudence, when the younger man would have preferred to embark on the life of the solitary. It is significant that when the *Annals of Ulster* recorded the deaths of eighty prominent celibates in the seventh century, only two of them were anchorites; and the seventh century may have been the period when the hermitage was at its most popular among Irish Christians. Temporary withdrawal from community, to spend a few weeks or even months alone in prayer, seems always to have been a more attractive option.

Nevertheless, there were Irish hermits, both at home and abroad. One of the peregrini who made his way to Germany was Marianus

Scotus of Mainz, who said his mass daily standing on the grave of his predecessor, with his own grave opened beside him. When St. Doulagh announced his intention to live as a hermit beside a church at Malahide (County Dublin) at the beginning of the seventh century, he set down his plan for the anchor-hold and how he and any successor would dwell in it. It was to be built of stone, twelve feet long and twelve broad, with three windows: one facing the choir through which the hermit could receive the Bread of Christ, another on the opposite side through which he could receive his food, and a third for light. The window for food was to be secured by a bolt and have an opaquely glazed lattice, which could be opened and shut, because no one should be able to look in, nor should the anchorite have a view out. The hermit was to be provided with nothing but a jar, a towel and a cup. After Terce he was to lay the jar and the cup outside the window and then close it. About dinner time he was to see if his food was there and, if it was, he was to sit at the window and eat and drink. When he had finished, whatever remained was to be left outside for anyone who might choose to remove it, and he was to take no thought for the morrow. But if it should happen that there was nothing for his dinner he was not to omit his accustomed thanks to God, even though he went hungry until the next day. His garments were to be a gown and a cap which he was to wear, waking and sleeping. In winter he might, if the weather was severe, wear a woolly cloak, because he was not allowed to have any fire, except what his candle produced.

As interesting as anything in that proposal is its reference to glazing. Doulagh's cell must have been one of the first Irish buildings to be fitted with an opaque form of glass to let light through, though glass beads have been found on the site of several ring forts in Ulster that have existed since the Iron Age or earlier. Although there is no evidence yet for the manufacture of enamel or glass from their raw materials in early medieval Ireland, there is lots of evidence for the recycling of glass. Scrap glass or cullet has been found at several sites where glass has been produced. The pieces are very small fragments of vessels such as beakers and palm-cups, originally made somewhere in Britannia, northern Gaul, Belgium or along the Rhine. If Doulagh's cell was indeed constructed as he proposed, this would not only have been a very early example of the Irish glazier's craft; it would also have been somewhat outside a tradition which was at pains to ensure that the eremitical life shrank from all conveniences.

13
The Curach

The building of Fionán's curach on the banks of the Shannon was in accordance with a tradition that was already several hundred years old. Open boats with hulls made of skin, intended for use at sea as well as on inland water, were certainly not peculiar to Ireland or to Wales, whose corwgl (coracle) was the local equivalent of the curach. The oldest reliable evidence for any such vessel in northern Europe exists in Norway, where a number of Bronze Age rock carvings at both Ostfold and Telemark clearly show long boats with wooden frames covered in hide. Of similar antiquity is the metal bowl found at Caergwrle, Wales, which is shaped like a boat, with detailed decoration suggesting that it was probably modelled on a skin-hulled vessel. An even more remarkable trophy is the boat model made from gold sheet that was found near Limavady, County Londonderry, in the nineteenth century, complete with fourteen oars, a steering sweep, a mast and a grappling iron, with marks on the hull that are clearly meant to depict the pressure points of a wooden frame upon panels of ox-hide. It is thought to have been made in the first century BC, and was discovered at a site only a mile or so from the River Roe, which would have been navigable to Lough Foyle, whence there is an open passage to the sea. Boats like this were used in all voyages made from Ireland before the ninth century: that much is certain, because the seagoing vessel constructed wholly of wood was unknown to the Irish before the coming of the Vikings, who introduced this novelty. Wooden dugouts for inland fishing and transport were another matter, almost certainly the very first craft that the Irish, among others, put into water.

There is an oral tradition that the first Irish to sail overseas did so about 222, when a fleet of curachs went raiding the coasts of Wales, Cornwall and possibly Gaul. It is impossible to say when, and in which direction, Irish seafaring went further afield than that, though a strong claimant must be Galicia, whose own skin-hulled ships had been coming to Ireland even before those first curach raids occurred. Trade was what drew the two Celtic regions together, however, and this would be a logical consequence of the legendary settlement of Ireland by people from Spain who arrived before the Fir Bolg, if there is any historical truth in that. Though we have almost no knowledge of what commodities were trafficked, there is plenty of evidence for exchanges between Galicia and Ireland once Christianity was settled in both lands, and it is chiefly of a liturgical nature. Contact was, however, broken when the Arabs invaded and then colonised Spain early in the eighth century. It is probably safe to assume that at no time did the Irish wolfhound, greatly prized in imperial Rome, travel by this route. Much more likely was its movement overland across Europe after being shipped to Gaul, with which there was a vigorous trade from at least the fourth century, wine, silk and iron going northwards in exchange for Irish hides, wool and hawks as well as hounds. Nantes, on the Loire, became the principal Gallic port for the Irish trade. It was there that Patrick's kidnapper, Niall of the Nine Hostages, was eventually killed on another of his freebooting expeditions.

We have to rely on Brendan's *Navigatio* for details of boatbuilding at this time. The curach that he constructed on Brandon Bay, big enough for at least fourteen men and their stores, had three layers of ox-hide, with two air chambers between them, stretched over the wooden frame, presumably as an insurance against the craft being holed, but perhaps also for insulation against an extremely cold sea, when the crew would expect to spend many days and nights on board. The hides had been tanned with oak bark and, after being stitched together with leather cords, the joints were sealed with holly resin, while fat was smeared on both sides of the hull to make it more surely watertight. The triangular lugsail, however, was made from the pelts of smaller animals because it had to be furled and stowed from time to time: ox-hide quickly lost its suppleness in seawater, whereas the thinner skins of wolf, pine marten and fox were relatively unaffected even when soaked with spray. It is unclear whether the mast was unshipped when not in use, or whether it was simply lowered on its step. The gunwales carried thole pins in which the oars were slotted, with probably one man pulling on each. But these were straight lengths of timber, without any form of blade; except for the steering oar, which was shaped like a spoon. We know virtually nothing more than this, except that Brendan and his men took with them enough hides to make two other boats, and the fat that would be needed for their preparation, together with food and a few other necessities for the crew.

15
Fasting

The origin of the forty-day fast that Brendan kept before embarking on his great voyage lay, of course, in their determination to imitate the forty days that Christ spent in the wilderness, tempted by the Devil. But fasting in general was already deeply rooted in Judaism, and rigorously practised by all devout Jews. The Desert Fathers subsequently regarded it as an article of faith, and performed prodigies of self-denial if all the stories told about them are to be believed. Antony is supposed never to have taken food more than every other day at best, sometimes every four days; Simeon Stylites is reputed to have gone from start to finish of Lent without anything at all as a matter of course; Macarius of Alexandria to have eaten nothing but cabbage leaves for the period before Easter. At Tabennisi, Pachomius ordained that cooked food was forbidden during Lent and that, during every week of the year, Wednesday and Friday must be days of total abstinence; the first of these fasts commemorating the day when Judas undertook to betray Christ to the chief priests, the second when Christ was crucified. Even this level of rigour was surpassed by St Jerome, who advocated fasting throughout the year, more severely still in Lent, less so in Pentecost. John Cassian, as always in favour of moderation, recommended to the Western Church that fasts should never last more than five days in one week, and that differences of sex, age and infirmity should always be taken into consideration when determining the length of any ascetical exercise.

There is little evidence that his advice was heeded in the early Irish Church, which approached the subject as zealously as anyone, the greatest enthusiasm being shown in the houses of religious. For the monks and nuns of Ireland there were three kinds of fasting, one of which was simply regular abstention from all food as a form of individual self-discipline, another the rejection of any food that was cooked, or in other ways made more appetising than in its raw state. Beyond this, there were the communal deprivations, usually observed in general throughout the Church. The most notorious of these were the great lenten fasts, of which there were eventually three in every year, all lasting forty days. From the outset of Christianity in Ireland, it was the custom to go short of food throughout the Lent preceding Easter, but the other two great periods of self-denial did not appear until early in the seventh century. One was known among the Irish as the Lent of Elias, and corresponded with the season of Advent leading up to Christmas; the other as the Lent of Moses, and was kept for the forty days of summer which followed Whitsuntide (Elias/Elijah and Moses were notorious ascetics in the Jewish tradition). Total abstinence on all Wednesdays and Fridays also became habitual, to the extent that Thursday was eventually known as dé dardóin, the day between the fasts.

It was in the individual regimes, however, that the Irish were at their most extravagant. When Columban arrived at his first Gallic monastery, Annegray, one of his brethren became ill. Having nothing to treat him with, the saint and the other monks stopped eating for three days and prayed for his recovery instead; and it worked, according to the hagiography. Four examples from Ireland suggest the range of self-denial adopted there. Maedóc, a monk of Ferns (County Wexford), neither ate nor drank anything for a full Lent, whereas the anchorite Moling fasted every day until sunset, except Sunday, unless someone visited him, in which case he would break bread with them at midday as well. The Rule that Comgall imposed on his monks at Bangor was so severe that seven of them died of hunger and cold. It is also recorded that at an unidentified Irish monastery, four monks once literally starved themselves to death, possibly in order to reach heaven and be with Christ earlier than would have happened naturally. Quite apart from their religious observances, the secular Irish also used fasting aggressively, as a means of coercing someone else into yielding a position or even some material advantage. If a man obtained a judgement against a debtor, and could not extract what was owed, he would sit outside his adversary's house and fast, in order to shame the householder, increase the debt and his own compensation. This form of blackmail is said to have come to Ireland from Egypt, but it was also a common pressure traditionally applied by the Hindu Brahmins, which both the Portuguese and the British noticed when they reached India (it may also, of course, be seen as a precedent for the twentieth-century hunger-strike in both Ireland and South Asia).

And it is not as though the Irish monks were living it up when they were fully fed; at the best of times, they were on exceedingly short commons compared with the Benedictines and other monastic orders which originated on the continent. Columban was of the opinion that food should be of poor quality and should never be eaten before the ninth hour (three o'clock in the afternoon), its purpose being to do no more than sustain life. At one time during his stay in the Vosges, he and his brethren ate nothing but bark and herbs, and when he took himself off on retreat in some anchor-hold, as he did periodically, he lived off blackberries and water. Under normal circumstances, the diet in his monastery consisted of whatever vegetables were locally available, together with bread, and permission had to be sought of the abbot for anything other than this to be consumed. Only on Sundays, feast days and the season of Easter was more and better food normally allowed, though if guests were staying at the monastery, the usual restrictions were relaxed not only for them but for everyone else. At Columcille's monastery on Iona, the regime was much more lenient, including milk and, when grain and vegetables ran out because there was a limit to what could be grown on such a small island, fish and the meat of both seals and sheep.

There was a certain amount of variation in the homeland of both these founding fathers. At the monastery of Cluain Eidnech (County Laois), only vegetables were eaten, and at Bangor (County Down) these were initially supplemented by bread alone, though milk and cheese were later admitted, too, after Comgall's time. Fish was more often than not regarded as acceptable where it could be freshly caught, and it is said that few houses objected to the consumption of game, on the grounds that it was less likely to excite the other bodily appetites than the flesh of cattle, pigs or sheep. Oddly enough, some monasteries appear to have found drinking beer to be unexceptionable—it was even taken in the rigorous Columban's continental monasteries—and a monk who touched nothing but water for thirty years was regarded as extremely hard on himself. But the relationship between good and plentiful food and intoxicating drink on the one hand, and sexual arousal on the other, was evidently a powerful reason why the Irish monastic diets were so severe. And not only in Ireland: Jerome, who was a Dalmatian and spent most of his life in Rome and the Near East, was adamant that liberality was bound to end in lust, and believed that bread, herbs and water were the most appropriate sustenance for monks. On the whole, the Irish tended to agree with him.

What these monks were giving up may be judged from the food that was generally available in Ireland between the fifth and thirteenth centuries. Cattle were kept for their milk or haulage capacities rather than for their meat; sheep and goats principally for their wool and hair; and only pigs seem always to have been raised with a view to eating them. The rabbit was not introduced until the Norman period, though hares existed much earlier. Domestic fowl may not have come to Ireland until the seventh century and, even then and for some time to come, neither they nor their eggs appear to have been taken very seriously as food, unlike the eggs of seabirds, which were commonly collected in coastal areas. Bees were kept extensively, honey being the only sweetener. Everywhere, cereals formed a large part of the diet in the shape of bread or porridge, and were harvested with sickles just below the ear so that the straw could be grazed by the animals. Other crops included onions, celery, various greens, peas, beans, possibly parsnips and carrots. Apples may have been cultivated, and there was an abundance of wild fruits—strawberries, raspberries, whortleberries, cranberries and rowanberries—as well as hazelnuts and acorns, which were highly regarded as food. Dulse, an edible seaweed, was eaten where it was available. Of fish, the Irish waters teemed with cod, haddock, plaice, flounder, salmon, salmon trout, eel, saithe, pollock, whiting, wrasse, while winkles, cockles and mussels were much sought after in the rock pools; but whelks, oysters and limpets were seldom harvested.

The limitations of space and generally shallow depths of soil, as well as distance from the nearest land, obviously had their effect on the diet of those offshore monasteries that were established from early in the sixth century. We can only guess how the monks of most island communities fared, but we have a better idea of the food available to the men who settled on Church Island, in the lee of the much larger Valentia Island (County Kerry) in the seventh century, because their midden has been excavated. It yielded various seashells, together with the bones of gannet, shag, cormorant, duck and white-fronted goose; also of seal, ox, pig, sheep, goat and horse, though whether the monks ate all the quadrupeds can only be a matter of conjecture. There was evidence of a small garden, but it seems unlikely that the community would have grown enough grain in it to keep them in bread, because cereals need a great deal more space for cultivation than the island could afford. The monks could without difficulty, however, have obtained all they needed from good arable land on Valentia Island or the Kerry mainland, both of which were very close.

15
Monastic Dress

The grey robes of the monks who lined the banks of the Shannon when Fionán and his companions left Clonfert in search of their anchor-hold were typical of Irish monastic dress at that time. This was invariably the same natural colour as the fleece of the sheep from which the wool was spun, as was the case with the French Cistercian order, the great European sheep farmers of the Middle Ages, but unlike the Benedictines, whose robes were dyed black. The Irish religious evidently chose not to colour their garments artificially, though this was within their power because plants providing dyes were all grown in Ireland at this time, and were made up into colourful textiles for people of means. Flax was also cultivated, so linen as well as woollen cloth was available.

The monks had, in fact, modelled themselves on the Desert Fathers in their clothing as in so many other respects, while making some allowance for the difference between their own climate and that of Egypt. Next to his body a monk wore the full-length tunica or léine, which was pulled on over the head like a smock, had sleeves and a shallow V neck: it was generally made of linen and, as it was provided with a girdle, it could be hitched up above the knees so that it didn't get in the way when manual labour had to be done; and there may have been a form of loin cloth underneath. Above the léine was a casula or culebadh, a woollen garment like a cape, woven coarsely and provided with a hood. If the weather was extremely cold, or if a journey had to be made, an amphibalus or brat would be thrown over everything else, like a shawl or a cloak. As with the Desert Fathers, Irish monks when travelling wore sandals and were each provided with a staff, crooked at the top. They would exchange these sticks if they met on a journey, as a mark of affection and brotherhood.

17
The Gallican Creed

The prayer that Fionán and the others began to recite while they were still at sea, and which continues to its end throughout the narrative of Part One, is an amended version of the Old Roman Creed, derived from the Gallican Liturgy. We do not know why the rites of northern Italy, Gaul, Spain and Ireland differed from the practice of Rome at an early stage in the development of the Church in these areas, but one theory maintains that the amended form was, in fact, the Creed used at Ephesus in apostolic times and therefore had the sounder pedigree. The shorter version may have been used in Rome as early as 150, though the first documentation for it is in the Apostolic Tradition of St Hyppolitus, who died in 235; that is, ninety years before the Council of Nicaea, which promulgated the Nicene Creed in order to defend orthodoxy against the Arian heresy. As a fundamental statement of Christian belief, this was adopted by both the Western and the Eastern (Orthodox) Churches and is still used by them today. In the Roman Catholic Church, the Nicene Creed is said at mass on Sundays, the greater feasts and the Feasts of Doctors, in the

Anglican Church at all communion services; otherwise the shorter Apostles' Creed is generally used.

Although so much is uncertain about the origin of the Creed that the Irish monks recited, one thing is clear: unless it reached Ireland before 432, it must have been brought by Patrick himself, who would have become familiar with it during his time in Lérins. Here it is in full, with the Gallic amendments to the Old Roman Creed set in italic:

Credo in Deum Patrem omnipotentem,
I believe in God the Father almighty,
creatorem caeli et terrae;
maker of heaven and earth;
Et in Jesum Christum, filium eius unicum, Dominum nostrum;
and in Jesus Christ, his only son our Lord;
Qui *conceptus* est de Spiritu sancto, natus est Maria virgine;
Who was conceived by the Holy Spirit, born of the Virgin Mary,
Passus sub Pontius Pilato, crucifixus, *mortuus,* et sepultus;
Suffered under Pontius Pilate, was crucified, dead and buried;
Descendit ad inferna;
He descended into hell;
Tertia die resurrexit a mortuis;
The third day he rose from the dead
Ascendit ad caelos;
He ascended into heaven
Sedit ad dexteram *Dei* Patris *omnipotentis;*
To sit at the right hand of God the Father Almighty,
Inde venturus est iudicare vivos et mortuos.
From thence he shall come to judge the living and the dead.
Credo in Spiritum sanctum,
I believe in the Holy Spirit,
Sanctam Ecclesiam *catholicam,*
The holy catholic church

Sanctorum communionem,
The communion of saints
Remissionem peccatorum,
The forgiveness of sins
Carnis resurrectionem,
The resurrection of the body,
vitam aeternam. Amen.
The life everlasting. Amen.

17
Skellig Michael

There can be no doubt that the rock which Fionán decided had been providentially revealed to him and his monks as the place they had been seeking, was Skellig Michael, which rises sheer from the Atlantic some eight and a half miles off Bolus Head in County Kerry. No other island with monastic remains off the west coast tallies with the circumstantial evidence. The monks clearly sailed south when they emerged from the River Shannon, after the wind slewed the bows of their curach in that direction. This rules out the offshore anchor-holds that lie north of the Shannon estuary. Some little distance in that direction is Inishmore, largest of the Aran Islands, where St Enda established an extremely rigorous monastery as early as 490 or thereabouts, at which those other monastic founders, Ciarán of Clonmacnois and Finnian of Moville, studied in their youth. It suffered from Viking raids, like everywhere else along that coast, and its last recorded abbot died in 1400. Further up, off the shores of Westmeath, St Rioch formed his community on Inishbofin about 530 and, apart from the Norse incursions, this monastery had the additional misfortune of being pillaged by raiders from Munster in 1015 and again in 1089.

County Mayo can claim two island monasteries: one of them, In-
ishglora, was started by Brendan in the middle of the sixth century,
and remains of the beehive huts in which his monks lived are still to
be seen there. Little is known about the foundation on Inishkea,
though the remains of huts have also been found there, together with
an early cross-slab with a Crucifixion carved on it, and a large quantity of purpura (dog whelk) shells, which suggests that the monks had
a flourishing scriptorium. Of the northern islands, none is more rewarding than Inishmurray, four miles into the Atlantic from County
Sligo, founded by St Molaise early in the sixth century and plundered
by Vikings in 802. But a great deal of the original structure has survived both the pillage and the passage of time, including the remains
of three churches, the enclosure wall, fifteen open-air altars, and many
cross-slabs. Most northerly of all these foundations, Columcille's
legacy from the sixth century, before he left Ireland for Iona, is on
Tory Island off the Donegal coast, which is unusual because the
monastic remains include a round tower, a comparative rarity in this
part of the country.

South of the Shannon are the remains of beehive cells and oratories
on Inishtooskert and Inishvickillane, which are two of the Blasket Islands off Kerry's Dingle Peninsula, but these can be discounted because
they are much closer inshore than the traditional description of Fionán's
skeilic, and they do not rise from the sea with the great drama of his discovery. For the same reason, the monastery on Church Island can also
be brushed aside. The island does not tower over its surrounding water,
and it is sheltered by Valentia Island, which is itself so close to the Kerry
mainland that they are attached to each other by a bridge.

Everything points to Skellig Michael, which answers all the de-
scriptions, not only in its relative isolation offshore, which is emphasised rather than diminished by the Little Skellig one and a half miles
away, but also in its own wealth of monastic remains; above all, by its
uniquely sensational appearance, an almost barren rock rising to a
pinnacle 715 ft above the ocean. It is not a very large rock, being less

than half a mile long, and at no point 500 yards wide, its stunning impact simply the result of its shape and its position, together with the knowledge that a community of monks lived on it for several hundred years of the Middle Ages. Very little of that history, in fact, has been documented: fewer than half a dozen dates, and nothing but the scantiest information associated with any of them. All other conclusions about Skellig Michael can be no more than informed guesswork, based on a study of the ground, and what is known about the early Irish Church generally. From the archaeological remains, for instance, the first settlement seems to have happened sometime between the latter part of the sixth century and the early years of the seventh: we can be no more certain than that, because the event went unrecorded. The name, however, is easily elucidated. Skeilic is simply Irish for a rock, especially a steep rock, and this one has also been variously spelt sgelic, scelig, scelec, sgeillac, sceilic, sceilg, sgelig, skeilig and scilig as well as skellig. The dedication to the archangel Michael (Mhichíl) occurred sometime between the middle of the tenth and the middle of the eleventh centuries, when he had come to be widely regarded as the patron saint of all religious communities situated on mountain tops and other high places, similarly associated with Monte Gargano in Italy, Mont St Michel in France and St Michael's Mount in England.

18
Nature Worship

The Celtic Church came closest to pantheism in the awe and gratitude that individuals still felt for the beneficent power of the sun, an instinct they had inherited from their pagan ancestors and amalgamated with their Christianity. The incantation of Fionán and the other monks to the sun which had revealed to them their skeilic, would therefore have come quite as naturally to them as their recitation of the Creed earlier or the Lord's Prayer at any time. The words uttered have been transposed to Ireland—though they, or something very close to them, might very well have been familiar to the sixth-century Irish—from Scotland, where they originated in a deeply traditional Celtic view of the natural world. They were first set down towards the end of the nineteenth century by Alexander Carmichael, exciseman and connoisseur of folk lore, whose greater life's work was the collection of such verse and prose during his travels around the Highlands and Islands. All had been transmitted orally for many generations but, as a result of Carmichael's interest, they were eventually published in six volumes as *Carmina Gadelica*. Carmichael had heard the morning hymn to the sun chanted by a very old man on South

Uist and by another on Mingulay, one of the outer isles of Barra. One of the reciters explained to him that 'There was a man in Arisaig, and he was extremely old, and he would make adoration to the sun and to the moon and to the stars. When the sun would rise on the tops of the peaks he would put off his head-covering and he would bow down his head, giving glory to the great God of life for the glory of the sun and for the goodness of its light to the children of men and to the animals of the world.' This is the sun prayer in full:

> *The eye of the great God,*
> *The eye of the God of glory,*
> *The eye of the King of hosts,*
> *The eye of the King of the living,*
> *Pouring upon us*
> *At each time and season,*
> *Pouring upon us*
> *Gently and generously.*
>
> *Glory to thee,*
> *Thou glorious sun.*
> *Glory to thee, thou sun,*
> *Face of the God of life.*

19

The World View

Bron's notion of the universe would have been typical of a fairly well-educated Christian towards the end of the seventh century, involving a tension between what astronomers and other intellectual explorers had surmised, and the teaching of a Church which needed to defend biblical accounts of creation that actually contradict each other in Genesis, with a timespan of about four thousand years between the creation and the life of Christ. The primitive view of the earth held that it was a disc with uplands, rivers and seas varying the flatness of its surface, with its boundaries encircled by an ocean that eventually reached a junction with the sky, the earth effectively floating like a vessel on waters that came from the floor of the universe. This was Homer's vision in the eighth century BC, whereas his near-contemporary Hesiod thought of the disc being suspended between the sky and the infernal regions; but from about 625 BC, the Ionian philosophers took note of the fact that the sun and the stars reappeared in the east after disappearing in the west, and gradually moved to the conclusion that the earth was a sphere rather than a disc. Pythagoras (c. 525 BC) was the first to suggest this and Aristotle demonstrated it some two hundred years later. He did so by using the same observations that have been put before children for the past several hundred years: the disappearance of a ship over the horizon and the circular shadow cast by the earth on the moon during an eclipse. Because Judaism and Christianity found this difficult to reconcile with Genesis, there was general scepticism throughout the western world until the sixteenth century, when Magellan led an expedition from Portugal across the Atlantic, down the coast of South America and across the Pacific to the Philippines, where he was killed in a skirmish with natives. But one ship out of his squadron continued the westward passage and reached Europe again in 1522, two years after Magellan set out, having completed the first circumnavigation of the world, and proved that the earth was indeed a globe suspended in air, and part of a wider universe.

21

The Buildings

The most remarkable feature of Skellig Michael, apart from the dramatic impact of the rock itself, consists in its surviving medieval buildings. These have been somewhat restored and uniquely exemplify the earliest semi-eremitical Irish religious communities, here arranged within enclosing walls on two terraces between 550 ft and 600 ft above sea level. The monastery's basis was the clochán (pl. clocháin) of thick drystone walls, shaped like a domed beehive on the outside, but roughly rectangular within, without window openings, but with low and narrow entrances. Such freestanding structures were peculiar to the most barren part of Ireland, though at Newgrange in County Meath the central chamber of a passage grave inside a mound was similarly shaped *c.* 2500 BC; as was the Greek Treasury of Atreus at Mycenae, another underground tomb going back to 1400 BC. All these buildings were distinctively constructed using a technique known as corbelling, in which the rising courses of stone each project slightly inwards until they form a dome, which is saved from collapse by the converging and carefully balanced pressure of masonry at the apex. The technical problem of the dome, and how big it could

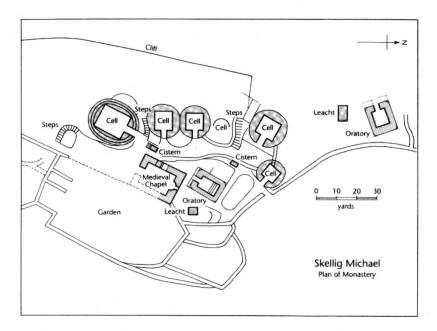

Skellig Michael
Plan of Monastery

be without disintegrating, was taxing builders in Constantinople at approximately the same time as the monks were making their dwellings on Skellig Michael. Haghia Sophia, the Church of the Divine Wisdom in the Byzantine capital, was reconstructed after a fire in 532 and a large dome was then introduced over the crossing, but collapsed some years later as a result of the lateral stresses. Further attempts had the same results (helped, very often, by the city's exposure to earth tremors) until a shallower dome was contrived in the fourteenth century, which we still admire as one of the architectural wonders of the world.

On Skellig Michael there were apparently six corbelled huts, varying in height between 16 ft 6 ins and 10 ft, with internal floor areas anything between 15 ft × 12 ft 6 ins and 8 ft 2 ins × 9 ft. Adjacent to them were two oratories, again with uncemented walls, but rectangular overall, with roofs curved and ridged to resemble the keel of a boat. These are smaller versions of the most famous of all early Irish medieval buildings, the Gallerus Oratory, which lies to the north of

Skellig Michael, on the Dingle Peninsula. The most dilapidated of all the structures on the skeilic now is the much younger—by perhaps 600 years—chapel, which has entirely lost its roof and much of its walls. The upper terrace also accommodates two leachta (sing. leacht), such as occur on all the offshore monastic sites down the Irish west coast. These are simply rectangular drystone constructions, which served various purposes: altar, prayer station, conceivably the tomb of an especially venerated individual. Human bones have been found close to one of the Skellig Michael leachta.

The buildings, the terraces, and the three different staircases that originally led from the monastery to landing places on the northern, southern and eastern sides of Skellig Michael, were all constructed from stone that either lay loosely exposed on the skeilic or was quarried there. The work was obviously done by the resident monks, and their only tools would have been iron hammers, chisels and crowbars, with leather bags, ropes and a primitive form of pulley for hauling heavy pieces of masonry. Other implements at their disposal would have been wooden or metal spades for tilling their garden, iron hooks and nets with stone sinkers for fishing. They might have had vessels of wood, earthenware or metal for food, but there would have been no cooking on Skellig Michael. Apart from the fact that this would have been contrary to the austere philosophy behind the deliberately isolated monastic life there, there was no fuel. Not so much as a bush, let alone a tree, has ever grown on the skeilic, and its limited deposits of soil are far too shallow ever to have provided turfs for burning.

23
Soul-friends

The word anamchara (literally meaning 'soul-friend') signified an early development in Irish monasticism, before the introduction of the novice master when the continental religious orders arrived in the twelfth century. The initial expectation that the young monk would learn by imitation also meant that a senior member of the community was assigned to keep an eye on him, and to dispense anamchairdeas, or spiritual direction. In short, he became a father in God to the novice; he also became his confessor, as the word has been traditionally understood. The responsibility this placed upon the older man was regarded as very grave, as this reflection from the ninth-century *Rule of Tallaght* makes plain. It refers to Mael Ruain, who founded the reforming Céli Dé movement. 'He used to say that soul-friendship was a parlous duty for, if one imposes on a man the penance that his sins have deserved, he is more likely to break it than to perform it. If the confessor does not impose the penance on him, that man's debts fall on him. "There are people," said he, "who think it penance enough just to confess . . ." '

It has been argued that it was from this relationship that the habit of regular and private confession spread outwards from the Celtic Church to the rest of Christendom, replacing the public confession which had been enjoined from at least the third century and generally regarded as something so weighty that it could only happen once in a lifetime. In fact, as in so many things associated with the early Irish Church, anamchairdeas and regular confession seems to have been proposed by John Cassian who, in Conference Two, after an anecdote about a penitent and the old man who advised him, remarks that 'if one does not confess a diabolic idea or thought to some soothsayer, to some spiritual person well used to finding in the magic, all-powerful words of Scripture an immediate cure for these serpent bites and the means of driving the fatal poison from the heart, there can be no help for the one who is in danger and about to perish.'

Now consider this. In 1783 Sir William Jones, whose pleasure was language and who had been well schooled in Classics as a boy, sailed for Calcutta to take up his post as a judge in the East India Company's colony there. In his leisure he began to study Sanskrit, and made a remarkable discovery: 'You would be astonished,' he wrote to a friend sometime after his arrival in Bengal, 'at the resemblance between that language and Greek and Latin.' He meant in matters of linguistic structure most of all, though there were other things that also chimed. He thereafter translated a number of Sanskrit texts into English, and propounded the quite startling theory that the Indo-European languages (which Jones carefully identified, excluding the Dravidian tongues of South India, likewise Magyar and the strange speech of the Basques) came from a common root; from which, long years after the judge's time, came the equally arresting idea (which Professor Renfrew and others still ponder today) that at some prehistoric time there was a simultaneous movement of population from somewhere in the region of the Caspian Sea, down into South Asia and westwards into Europe. It is therefore not quite as unexpected as it might otherwise be, to hear an echo of anamchara in the Sanskrit word Brahmacharya, which describes the course of discipline and education accepted by every student of Hinduism under the direction of a guru. This is, after all, a much less remarkable correspondence that the one between the Attic Greek word for fish (ikhthus) and the Maori usage (ika): in Maori mythology, the North Island of New Zealand is Te Ika a Maui—the fish of Maui.

23
The Monastic Horarium

The division of day and night into regular prayer times is as old as monasticism itself, though Cassian was exasperated to find that almost every community he visited throughout the Near East had its own idiosyncratic schedule. There seems also to have been a certain amount of variety from the start of the Celtic Church. The oldest timetable in Irish, set down in the middle of the fifth century, refers to six offices, and this is believed to have been a norm throughout the sixth century, too. The offices were: Teirt (Terce, at 9 a.m.), Medón lái (Sext, at noon), Nóin (None, at 3 p.m.), Espartu (Vespers, at 6 p.m.), Midnocht (Nocturns, at midnight), Iamérge (Matins, at 3 a.m.); and these times were invested with a very profound mystical significance. Terce was the hour when Christ was sent to Pilate, and the Holy Ghost descended upon the Apostles; Sext the hour of Adam's sin and the Crucifixion; None the time when Christ died and the angel visited the centurion Cornelius; Vespers when sacrifice under the Old Law (of the Jews) took place; Nocturns when God created the elements; Matins when Peter denied Christ for the third time, coinciding with Christ's sufferings in the house of Caiaphas. In the seventh century Prime was added to the original six offices, and took place at six in the morning, which had always been regarded biblically as the first hour of the day.

Taking their cue from the Desert Fathers, the Irish structured the offices round the recitation from memory of the entire Psalter every few days. At Prime, at Terce, at Sext and at None no more than three psalms might be sung, the most prolonged activity occurring at the night offices, when up to thirty-six had to be accounted for each time the monks assembled for Matins between Monday and Friday. On Saturday and Sunday this figure rose to as many as seventy-five, which meant that the office lasted for several hours and the monks got very little sleep. Also derived from Egypt was the practice of dividing a community into equal groups of up to four brethren, who took turns at the singing. This was done standing, while the listeners sat, though all rose for the choral recitation of the Gloria at the end of each psalm. Psalmody apart, there might be occasional hymns, and a lesson would be read from both the Old and the New Testaments, the latter augmented by an extra passage from one of the Gospels at the weekend. Numerous collects were introduced between the various psalms, and other prayers were said. Eventually these became formalised into an unvarying order; for the monks' own sins, for the whole Christian people, for all priests and clergy, for those who bestow alms, for peace among kings, for enemies. The practice of the Irish was most markedly different from that of the Desert Fathers in the frequency with which they abased themselves during the offices. After every psalm they bowed to the altar, and during the collects they went to their knees, but these were mere preliminaries. Their worship was also punctuated by kneeling with arms outstretched in the posture known as crosfigel, also by throwing themselves fully prostrate to the floor, and they clearly did this as a matter of individual devotion; an eighth-century monk named Oengus habitually made three hundred genuflections every night.

The mass, the central act of Christian worship, was celebrated every Sunday and on feast days, usually not long after dawn, but on the great feasts not until midday. It began with a confession of sins, followed by a litany of the saints and other prayers, together with the Epistle, the Gospel and the Credo, then the commemoration of the dead, before the consecration of the Host with a solemn expression of faith by the celebrant: 'We believe, O Lord, we believe that in this breaking of Thy Body and this pouring forth of Thy Blood we were redeemed; and we trust, strengthened as we are by the reception of this sacrament, that what we now possess in hope we shall enjoy in truth and enduringly when we reach the heavenly kingdom.' In their vernacular speech, the Irish commonly referred to the consecration as 'making Christ's body' on the altar. It was the custom for Holy Communion at the mass to be received in both kinds, but it is possible that although wine was imported from Gaul at an early stage, the Irish may often have used their equivalent beverage, mead, instead; and it is conceivable that a monastery as remote as the one on Skellig Michael, which wished to have as little as possible to do with the mainland its monks had deliberately turned their backs on, might well have celebrated with bread alone, in its earlier centuries at least. The most convincing argument in favour of a boat's coming out to the skeilic periodically is, in fact, that the community there could not have survived spiritually without at least bread for communion, which the monks could not provide out of their own resources.

We know next to nothing about the music of Irish worship during those early centuries. The *Antiphonary of Bangor,* composed at Columban's old monastery between 680 and 691, tells us a great deal about the metre of hymns and other chants used at that time, some of which were obviously grounded on Latin models, others being quite alien to Roman liturgies; but it does not include the musical notation. Nor is the *Stowe Missal,* written in the eighth century but based on earlier material of Roman, Gallican, Spanish and Irish origin, any more illuminating. To what extent the singing of the first Irish monks resembled the plainchant that became associated with Gregory the Great (though it owed more to the stimulus provided by the later Pope

Vitalian and the Emperor Charlemagne), we shall almost certainly never know. The earliest surviving manuscripts and treatises on plainsong appear to have been composed late in the eighth century in the monastery of St Gall, which the Irish monk who accompanied Columban to the continent founded in Switzerland.

31
Cú Chulainn

The legendary figure of Cú Chulainn, whom the monk Bron remembered well from boyhood, is one of the great characters in the Irish mythology. The saga of which he is the central character, *Táin Bó Cuailnge* (*The Cattle Raid of Cooley,* which is a peninsula in County Louth), was part of a purely oral tradition which was not committed to writing until, possibly, the start of the seventh century; but scholars have long recognised that in spite of many supernatural events it is related to some extent to the actual history of Ireland, and they differ only in their assessment of period, which has been put variously between the first and fourth centuries. It is set in the reign of a king of Ulster, Conchobar mac Nessa, and it is pitched wholly at the level of chiefs and princelings, most of whom are in the heroic mould. As the long epic unfolds, the occasions when women with friendly thighs submit to aggressively priapic males is exceeded only by the number of extremely violent and bloody acts. No one is more vigorous, in every sense, than the braggart Cú Chulainn, nephew of Conchobar and the apple of every Ulster eye, who spends much of the tale holding back an army of invaders from Connacht, performing valorous deeds time after time and suffering grievous wounds in the process. At one stage he kills 130 kings in a single bloodbath, and the climax of the epic is his fight with Fer Diad, his own foster-brother, which Cú Chulainn inevitably wins, after both men have behaved honourably.

There were many versions of the *Táin,* and many other associated stories. Here is a fragment of one about the death of King Conchobar, written by an unknown author, possibly in the ninth century. It is an excellent example of the Irish pagan mythology's becoming mingled with the new beliefs of Christianity, in the way that Bron on the skeilic recalled from his youth.

His head was healed then, and was sewn up with a golden thread, for the colour of Conchobar's hair was like the colour of gold. And the doctor told Conchobar he should take care that anger should not seize him, and that he should not mount on horseback, and should not have to do with a woman, and should not eat food gluttonously, and should not run. So he remained in that dangerous state, as long as he lived, for seven years, and he was not able to be active but to stay in his seat only; until he heard that Christ was crucified by the Jews. A great trembling came on the elements at that time, and heaven and earth shook with the monstrous nature of the deed which was done then—Jesus Christ the Son of the Living God crucified though guiltless. 'What is this?' said Conchobar to his druid. 'What great evil is being done today?' 'It is true,' said the druid. 'It is a great deed that is done there, Christ the Son of the Living God crucified by the Jews.' 'That is a great deed,' said Conchobar. 'That man,' said the druid, 'was born the same night that you were born, that is, on the eighth day before the calends of January, though the year was not the same.' Then Conchobar believed; and he was one of the two men in Ireland who believed in God before the coming of the Faith, and the other was Morann. 'Well now,' said Conchobar, 'a thousand armed men shall fall at my hand in rescuing Christ' . . .

32
Two Tonsures

As may be supposed from that extract, the golden-haired man was the archetypal Irish hero, pictured most extravagantly in the person of Cú Chulainn himself.

You would think he had three distinct heads of hair—brown at the base, blood-red in the middle, and a crown of golden yellow. This hair was settled strikingly into three coils on the cleft at the back of his head. Each long loose-flowing strand hung down in shining splendour over his shoulders, deep-gold and beautiful and fine as a thread of gold. A hundred neat red-gold curls shone darkly on his neck, and his head was covered with a hundred crimson threads matted with gems . . .

The antithesis of this shaggy effulgence, the totally shaven head, was the caste mark of the slave among the pagan and early Christian Irish, and therefore repugnant to all males. Even the highly esteemed druids could only bear to shave the front of their heads, leaving long hair flowing at the back, and the first Irish monks followed druidical example by shaving in front of a line drawn from ear to ear. This peculiarly Irish form of tonsure was to have a most disproportionate effect on Church history, after Gregory the Great despatched Augustine to Britannia in 596 on a great missionary enterprise that would, among other things, result in his becoming the first Archbishop of Canterbury.

Augustine was charged with two overriding objectives: to take Christianity to the pagan Saxons, and to correct some practices in the Celtic Church which Rome deemed unacceptable. The Church had been established in Britannia by missionaries from Gaul crossing the Channel not later than the end of the second century, being well enough established in 314 for three of its bishops—Eborius of York, Restitutus of London and Adelfius of Caerleon-on-Usk—to attend the Synod of Arles. It flourished until the great Saxon incursions in the second half of the fifth century pushed it out to the margins of the island. Augustine sent his man Paulinus to resurrect the faith in Northumbria on Roman lines, but success was short-lived (it had depended on the patronage of a king, who died within a few years) and Paulinus returned south to report failure.

The Celtic Church was subsequently revived in the North as a result of Irish influence, in a chain reaction which began when Columcille/Columba left Ireland for Scotland in 563 and presently established his own monastery on Ioua Insula (which would not be known as Iona until the fourteenth century). This was not the first mission to Scotland: that distinction belongs to the Cumbrian Ninian, who settled at Whitehorn in Galloway at the end of the fourth century, and had some influence on the early Irish Church. But it was from Iona in 635 that the monk Aidan went forth to Northumbria in response to a plea from its King Oswald, and founded a monastery on Lindisfarne which, like Iona, adopted the spiritual traditions and the mannerisms of the Irish. The stage was being set for another great conflict between Romans and Celts, comparable to the one that had seen the latter driven to the western extremities of Europe by the imperial legions. This time, Rome was not hungry for territory so much as religious authority. It still demanded total submission, though.

32
The Synod of Whitby

As matters came to a head, Roman demands focused on two Celtic usages in particular, one of which at this distance seems ludicrous. Submission meant that the Irish must abandon their distinctive tonsure and adopt the Roman style, which was a totally shaven crown with a circle of short hair round the sides of the head. The other requirement was much more serious and had a great deal to commend it. These two branches of the Western Church were out of step in celebrating Easter, but a paschal controversy had been dribbling on throughout Christendom for four hundred years. The earliest method of computing Easter was based on the Jewish formula for determining the Passover on a fixed day of the lunar month, which was governed by an 84-year cycle. But the Church in Alexandria, which led in scientific scholarship at that time, in the second century advocated an alternative cycle of nineteen years, based on the latest astronomical findings. There was yet a third variation of 112 years, which meant that in 387 Easter was observed in Gaul on 21 March, in Rome on 18 April, and in Alexandria on 25 April. Roman patience at last ran out in 463, when Pope Hilarus appointed Abbot Victorius of

Aquitaine to settle the matter once and for all, and he came up with a cycle of 532 years. The Celtic Church, however, persisted with the very earliest calculation, which Patrick had brought to them. This meant that as late as the middle of the seventh century in newly revived and mostly Celtic Northumbria, Queen Eanfleda, following the Roman prescription, was fasting on her Palm Sunday, while her husband King Oswy (Oswald's son) was celebrating his Easter with a great feast.

Under pressure from the Roman party, Oswy summoned both factions to Whitby in 664, to a synod which was held in the dual monastery there that Hilda, following Brigid's example in Kildare, had founded five years earlier. She sided with the Celtic churchmen, who were led by the Irishman Colmán of Lindisfarne, Bishop of Northumbria, and whose principal opponents were Wilfrid, the Roman-trained and Roman-tonsured Abbot of Ripon, and his patron Agilbert, Bishop of the West Saxons. By then, Colmán was fighting a rearguard action, for the Celtic Church in southern Ireland had already decided to accept the ruling of Rome, following a directive from Pope Honorius, who exhorted the Munstermen 'not to think their small number, placed on the utmost border of the earth, wiser than all the ancient and modern churches of Christ throughout the world.' Effectively, Colmán's only remaining allies outside Northumbria were the northern Irish and the men of Iona.

And after Oswy had heard the arguments from both sides, Northumbria was lost to Rome as well. Hilda accepted the king's decision and adjusted the life of her monks and nuns in Whitby, but Colmán retreated to Iona, which obdurately stuck to the old ways even when the northern Irish had accepted the inevitable: Columcille's old monastery did not celebrate Easter on the Roman date until 716, and clung to the Irish tonsure for a further two years. This was in spite of the fact that, as soon as the Synod of Whitby ended, Theodore, Archbishop of Canterbury, announced that 'those who have been ordained by bishops of the Irish or Britons, who are not

Catholic as regards Easter and the tonsure, are not deemed to be in communion with the Church, but must be confirmed by a fresh laying on of hands by a Catholic bishop.' Otherwise they were heretical and excommunicate.

43
Columcille and Iona

The obduracy of Iona can be traced to the sturdy character of its founder, Columcille, who belonged to the royal line of the Ui Niall; was, no less, a great-great-grandson of the Niall Naoi Ghiallach who had made the young St Patrick captive. Because he left no record of his life, there are inevitably question marks against the received accounts of Columcille's religious formation. Apart from the fact that he was born c. 521, very little is certainly known about him before he left Ireland in 563 other than that, according to custom, he was fostered on the priest Cruithnecán. His biographer Adomnán of Iona tells us he then studied under one Gemmán in Leinster, later under the hazy figure who is variously named Finnbarr, Finnio and Uinniau in Adomnán's book. This imprecision is at least partly responsible for the confusion that has ever since surrounded the circumstances leading up to Columcille's exile. Some commentators have thought it likely— but it is unproven—that Finnbarr was none other than the Finnian who founded a monastery at Moville in County Down, where Columcille was perhaps ordained deacon before passing into the hands of Finnian of Clonard, the equally illustrious 'teacher of saints', who nurtured him further and priested him. According to these accounts, Columcille then went to a monastery at Glasnevin, near Dublin, before returning to his own familiar lands in Derry, where he founded the first of several monasteries he would leave behind in Ireland, Durrow among them.

There are two traditional accounts of what happened next, and they do not necessarily contradict each other. The more vivid version associates Columcille with the Battle of Cúl Drebene, which did actually take place beneath the table top of Ben Bulben, County Sligo, in 561. This was old-fashioned Irish inter-tribal warfare between, in this case, the ard rí Diarmait mac Cerbaill and his kinsmen of the southern Ui Niall, and an alliance of Connachtmen and the northern Ui Niall. The allies could certainly have counted on Columcille's emotional support, because he was himself of the Ulster Ui Niall, though it is doubtful whether he took part in the fighting, as one source hints. It is much more plausible that he prayed for the success of his kinsmen, who triumphed at a cost of 3,000 enemy dead. To this story is attached the theory that he went overseas in expiation of his sin in causing such casualties with his prayers: forced into exile, according to some, but from guilt-ridden choice in other tales. The second legend about Columcille has him illicitly copying a Vulgate Psalter which belonged to his old mentor Finnian of Moville and, when he refused to give it up, being hauled before the high king, who judged that the illicit text must be returned to Finnian's library, uttering the memorable phrase, 'To every cow its calf, to every book its copy.' Again, Columcille goes into exile as a result of his fault.

The *Annals of the Four Masters,* which were not compiled until the seventeenth century, attempted to marry the two legends by suggesting that the Battle of Cúl Drebene occurred partly because of 'the false sentence which Diarmait passed against Columcille about a book of Finnen . . .' This can be nothing more than conjecture, when Adomnán is silent about the Psalter and mentions the warfare very casually: 'In the second year following the battle of Cúl Drebene, when he was forty-one, Columcille sailed away from Ireland to Britain, choosing to be a pilgrim for Christ.' Adomnán was not only writing much closer to the event—he was born *c.* 627 and died in 704—but he was a blood relative of Columcille and became the ninth Abbot of Iona after him. The weakness of his hagiography is that it is heavily overweighted in favour of miracles performed by the saint. But there can be no denying Columcille/Columba's strength of character or his importance in the development of the Celtic Church after the passing of Patrick. Whether he is remembered by his Irish name (which means 'Colum of the church', and which is today inscribed on stained glass in the Abbey on Iona) or its Latinised variant (which can also be read as 'dove'), it is a name that has rung almost as clearly as Patrick's down all the years since.

48
Mingled Traditions

The confusion in Cainnech's head, as he prepared to make up the quarrel with Macet in the old Irish fashion by kissing nipples, is yet another example of pagan and Christian traditions becoming mingled in the mind of the early medieval believer. His awareness of Christ's words at the Last Supper, and of the Te Deum, have merged with his memory of something from the *Táin* stories, when Cú Chulainn lusts after the beautiful Emer.

Cú Chulainn greeted the troop of girls and Emer lifted up her lovely face. She recognised Cú Chulainn, and said;
  'May your road be blessed!'
  'May the apple of your eye see only good', he said.
  Then they spoke together in riddles.
  Cú Chulainn caught sight of the girl's breasts over the top of her dress. 'I see a sweet country,' he said. 'I could rest my weapon there.'
  Emer answered him by saying:
  'No man will travel this country until he has killed a hundred men at every ford from Scenmenn ford on the River Ailbine, to Banchuing—the Woman Yoke that can hold a hundred—where the frothy Brea makes Fedelm leap.'
  'In that sweet country I'll rest my weapon', Cú Chulainn said . . .

49
Monastic Feuds

The abbot who punished Cainnech and Macet was not lightly dismayed by the belligerence of Irish monasteries in the eighth century. Several things help to explain this phenomenon, and one is the long tradition of violence among the Irish, including human sacrifice, of prisoners among others, which still flourished when Patrick arrived in the fifth century. The ancient Brehon law, moreover, insisted that injured parties must take their own revenge and not expect society to dispense justice on their behalf. Another factor is that in the seventh century there had been a tremendous growth of monastic property and monastic wealth, which made them tempting targets of plunder. A fourth thing is that, until 804, monks were not exempt from the military service that all men were obliged to give their tribal superiors in exchange for protection and other favours: they simply were not allowed to become lifelong pacifists after taking the tonsure. It may be shocking, but it was perhaps inevitable that these carefully nurtured aggressive instincts sometimes spilled over into disputes between rival monasteries.

The monastic battles are well enough documented. In 673 the monks of Clonmacnois fought the monks of Durrow, and two hundred of the Durrow men were slain. Other engagements took place between Clonmacnois and Birr in 760, between Cork and Clonfert in 807, between Kildare and Tallaght in 824. It is clear that battle was sometimes joined because the leader of one faction had simultaneous roles, as both abbot and tribal chief. The ninth-century Munsterman Feidhlimidh was just such a one, and he had a long record of aggression against other monasteries. He slew monks at Clonmacnois and burned their sanctuary in 833; three years later he stormed the abbey of Kildare, where the Archbishop of Armagh and his clergy had taken refuge; burned Armagh itself in 840 and, six years after that, attacked and plundered Clonmacnois once again. But Armagh could give as good as it got when opportunity arose. An eighth-century bishop of that see, angry because provisions had been stolen from some monastery under his jurisdiction, went to war with the King of Ulster, cornered him in battle at Faughart, and had him beheaded, even though he had taken refuge in the church.

50

Penance

The sacrament of penance did not become general in the Western Church until the Fourth Lateran Council, in 1215, required all Christians to confess and expiate their sins at least once every year. It had a much longer history than that, however, in the Irish Church, where penance was as much a characteristic of early monastic life as was the habit of peregrinatio, involving exile from the monk's native land. As in many other aspects of the life, this imitated patterns already established by the Desert Fathers in Egypt, where a similar form of ascesis was increasingly prominent from the third century onwards. But it has also been suggested that the Irish were readier for penance than other Europeans, in order to compensate for their lack of suffering when they were uniquely evangelised without bloodshed: penance, in short, was a psychological necessity which they welcomed in order to feel that they were true disciples of the suffering Christ. Patrick himself is associated with a document which was a foretaste of what was to come in Ireland later, by indicating areas of impermissible behaviour without specifying penalties as precisely as his successors would. In the *Canons* of a Synod of Patrick, Auxulius and Isernius, which were formulated sometime after 439, the following passages are typical:

1. If anyone for the redemption of a captive collected in the parish on his own authority, without permission, he deserves to be excommunicated . . . 10. If anyone has shown the beginning of a good work in psalm singing and has now ceased, and lets his hair grow, he is to be excluded from the Church unless he restores himself to his former condition . . . 17. A virgin who has taken the vow to God to remain chaste and afterwards marries a husband in the flesh shall be excommunicate until she is converted: if she has been converted and desists from her adultery she shall do penance, and thereafter they shall not dwell in the same house or the same village . . .

A later generation produced the first of the true penitentials, which were little more than tables of offences with the allotted penalty set down beside each, and which would be composed more prolifically by the Irish than by anyone else in Europe, where they continued to appear from a number of sources until the sixteenth century. One of the earliest such Irish documents may have come from the hand of Finnian of Clonard, mentor of Columba/Columcille, sometime between 525 and 550. It begins, 'In the name of the Father and of the Son and of the Holy Ghost. 1. If anyone has sinned in the thoughts of his heart and immediately repents, he shall beat his breast and seek pardon from God and make satisfaction, that he may be whole.'

Soon afterwards, the author moves from the vague to the precise.

6. If after anyone has started a quarrel and plotted in his heart to strike or kill his neighbour, if the offender is a cleric, he shall do penance for half a year with an allowance of bread and water and for a whole year abstain from wine and meats, and thus he will be reconciled to the altar. 7. But if he is a layman, he shall do penance for a week, since he is a man of this world and his guilt is lighter in this world and his reward less in the world to come.

There are fifty-three clauses in Finnian's *Penitential,* and they cover every one of the sins proscribed by Christianity, before concluding with, 'Here endeth this little work which Finnian adapted to the sons of his bowels, by occasion of affection or of religion, overflowing with the graces of Scripture, that by all means all the evil deeds of men might be destroyed.'

The most comprehensive of these documents was the *Penitential* of Cummean, an Irish abbot (*c.* 590–662), who founded a monastery at Kilcummin, County Offaly, and may also have been at Clonfert, though some commentators believe his latter years were spent in Columban's old Italian foundation at Bobbio: certainly his *Penitential* was widely circulated throughout the Frankish Empire by the early ninth century. Its 182 clauses are arranged in eleven sections which cover not only the statutory sins but also Petty Cases, a group which appears under the heading 'Let Us Now Set Forth the Determination of Our Fathers Before Us on the Misdemeanours of Boys' and, finally, a bundle of penances which arise out 'Of Questions Concerning the Host'. In the case of men who become drunk in their gluttony, 'if they have taken the vow of sanctity, they shall expiate the fault for forty days with bread and water; laymen, however, for seven days'. Punishments for avarice include this: 'He who hoards what is left over until the morrow through ignorance shall give these things to the poor, but if he does this through contempt of those who censure him, he shall be cured by alms and fasting according to the judgment of a priest. If, indeed, he persists in his avarice, he shall be sent away.' The angry man is told that 'He who refuses to be reconciled shall live on bread and water for as long a time as he has been implacable.' The proud run the risk that 'He who intentionally disdains to bow to any senior shall go without supper.' The petty offender who arrives late in choir 'shall sing eight psalms in order', while 'Boys talking alone and transgressing the regulations of elders, shall be corrected by three special fasts.'

In all the penitentials, the greatest attention was paid to sexual offences, leaving the impression that these were regarded with even more anxiety than violent crimes: it has been well said that 'with peculiar intimacy they (the penitentials) reveal the faults of men and of society in a far-off age, as well as the ideals of the monastic and ecclesiastical leaders on whom responsibility was laid for the guidance of souls'. Finnian is rather easy-going compared with what came afterwards: 'If any cleric lusts after a virgin or any woman in his heart but does not utter his wish with the lips, if he sins thus but once he ought to do penance for seven days with an allowance of bread and water.' Repetition invites a penance of forty days. Cummean was much more vigorous than that in his section 'Of Fornication', which runs to thirty-three clauses and considers separate penalties for every rank of cleric found guilty of different improprieties: sin committed with a beast, a man's defilement of his mother, sodomy, masturbation, fellatio and even nocturnal emissions—'He who desires to sin during sleep, or is unintentionally polluted, fifteen psalms, he who sins and is not polluted, twenty-four.'

The most common penance, whatever the individual's failure might have been, was to be restricted to bread and water for anything up to a year, even longer in exceptional cases. Other penalties included flogging (Columban was especially keen on that), payment made in slaves, the plucking out of hairs, extremely uncomfortable sleeping arrangements and, as an ultimate sanction, excommunication. Exile was occasionally imposed, with instances of someone being cast offshore in a small boat with a single paddle (in pagan Ireland, criminals had been cast adrift in boats). There is also the story of St Enda, who was exiled to Britannia for some offence and asked how long he should remain there. 'Until such time as the fame of your good deeds shall come to us', was the chilling response. But self-inflicted penances could be the most frightful of all, as in the case of Findchú of Brí gobann, who for seven years suspended himself from iron sickles hooked under his armpits.

The offence for which Cainnech and Macet were punished on their skeilic had its roots, as they protested, in the Irish tradition of kissing nipples in order to affirm friendship, especially after a quarrel.

St Patrick refers to it in his *Confessio,* when he is told in a dream to leave his Irish bondage and sail overseas. After a small altercation with the captain of the waiting ship, he is told, ' "Come, because we are receiving you on faith, make friendship with us in whatever way you will have wished"; and on that day, to be sure, I refused to suck their nipples on account of the fear of God . . .'

The penance imposed upon Cainnech and Macet embodied two of the more common features of monastic punishment, one of them virtually unheard of outside the Irish Church. The recitation of psalms as a penalty occurred throughout Christendom and, of these, Psalm 119 (known as Beati Immaculati from its opening verse 'Blessed are those that are undefiled in the way: and walk in the law of the Lord') was probably invoked more than any other, its 176 verses making it notoriously much longer than anything else in the Psalter. To sing the Beati while at the same time standing or kneeling in the posture known as crosfigel (cross-vigil), with arms outstretched from start to finish, was a peculiarly Irish form of punishment and must have been excruciating for only one round, let alone several. Interestingly, however, the idea of chanting sacred verses for lengthy periods in order to atone for some fault was not confined to Christians. The practice had also developed within the Brahminical tradition of India.

55
Sin

Obviously underlying the notion of penance was the Christian philosophy of sin. Much of this was, of course, inherited from Judaism, beginning with the disobedience of Adam and Eve, being subsequently codified in the Law of Moses and amplified by the Prophets. The authors of the New Testament variously emphasised the faults in an individual's character, the violation of natural law, personal responsibility, and simple disbelief in Christ. Beyond these fundamentals, the theologians of the very early Church saw the greatest dangers of all as threefold: idolatry, which meant a return to paganism; fornication and other sexual offences, and bloodshed. It was the archdeacon Evagrius Ponticus (346–99) who introduced the concept which was later to be popularised as the Seven Deadly Sins. A love affair had caused him to leave Constantinople for Jerusalem, before he moved on again to spend the rest of his life with the monks at Nitria. Among his subsequent writings in the Egyptian desert was a treatise *On the Eight Evil Thoughts,* which classified the basic sins as gluttony, fornication, avarice, dejection (tristitia), anger, weariness of spirit (accidie), vainglory and pride, and divided responsibility for them between the three parts of the soul which had been identified by Plato (the rational, the desiring and the spirited). It was Pope Gregory the Great who later added envy to the list, while amalgamating tristitia and accidie. Evagrius was one of the principal Christian thinkers in Egypt during John Cassian's time there, and Cassian absorbed much of his teaching in the formation of his own philosophy. Among other things, his *Institutes* contain an extensive description and elaboration of the sins catalogued by Evagrius, Cassian thus becoming the conduit through which they passed to a wider audience in western monasticism.

55
Poverty and Sickness

Below the level of the chieftains and princelings who continued to spawn heroic myths, life in ninth-century Ireland was as harshly precarious as it was elsewhere in Europe at that time. The fact that the earliest dwellings—the crannogs in the middle of lakes, and the ring forts whose foundations have survived in their thousands to this day—were still in use, and all paid careful attention to their defences, speaks eloquently of the potentially hazardous existence most Irish then endured. True, the arrival of Christianity had brought certain economic advantages, with a limited reduction in the tribal fighting and cattle-raiding, and the introduction of arable farming to augment husbandry. But even in more peaceful times there seems to have been little for later generations to envy.

For everyone above the level of the slaves and the hereditary serfs, the basic unit of currency was the value of a milch cow until the Vikings introduced coinage in the mid-ninth century. One such beast would buy twenty-four sacks of corn or malt, twenty-four cows would buy over thirty-four acres of the best land (whereas a similar area of turf bog could be obtained for only eight dry cows). So the loss of a milch cow as a piece of capital, quite apart from its milking potential, was considerable to anyone living off the land with a family to feed. Accidents like the one which befell Enda's family cow were all too common, as were veterinary diseases (murrain) of one sort and another. Arable farming, too, was a risky business in the Irish climate, as the various annals make plain when they keep track of bad weather and other rural calamities across the centuries:

Great drought this year . . . A mortality of cattle and birds, such that the sound of a blackbird or a thrush was scarcely heard that year . . . A year of scarcity and hunger . . . A great downpour and much of the corn crop was abandoned . . . A murrain of cattle and pigs . . . A violent wind and it greatly damaged the corn crop . . . A violent wind, which sundered the grain from the corn . . . In the above year the cattle were afflicted vastly with sinech, many lumps . . . Heavy snow this year, and a great loss of cows, sheep, and pigs the same year . . .

Nothing apart from violence, however, threatened human life in early medieval Ireland more than sickness. Tuberculosis, which had been known in Europe since at least 5000 BC, was almost certainly rife in the damp climate beside the eastern Atlantic, and leprosy had been gradually spreading northwestwards from Egypt, where it was endemic long before the time of Christ. Leprosy was a collective word in the Middle Ages (like plague) which covered all chronic skin diseases (true leprosy's particular microbe was not separately identified until 1873) and there was no treatment for any form of it: it either receded naturally and the victim appeared to have been miraculously cured by one of the saints, or it persisted, in which case he was universally regarded as unclean and required to live apart from the rest of society. Having come from the East with the returning Roman legions, the most disfiguring form of leprosy (now known as Hansen's Disease) had been transmitted to Gaul in the first century AD, and made its appearance in Ireland within another four hundred years or so. Even in the ninth century, a leper who appeared in an Irish village would have been fortunate to survive on local charity.

As for the various kinds of cancer, and other ailments causing great pain, they simply had to be borne, for western medicine had advanced no further than Galen took it in the second century. Bloodletting was the commonest form of treatment, and astrology stood high in every medical practitioner's craft, while the traditional herbal remedies that would have been known to every village elder had only a limited range and effect. They could do nothing where precise surgery might have been successful; but no real progress in surgery would be made until human dissection began at the University of Bologna in the fourteenth century. Arabic medicine was far ahead of European practice in the Middle Ages, but Christians were apprehensive of Islam by then and therefore loth to learn from it, as the testimony of a twelfth-century Muslim doctor makes plain.

He had been asked to see two Europeans who became sick during the disastrous Second Crusade to Mesopotamia:

They brought me a knight with an abscess in his leg, and a woman troubled by fever. I applied to the knight a little cataplasm; his abscess opened and took a favourable turn. As for the woman, I forbade her to eat certain foods and I lowered her temperature. I was there when a Frankish doctor arrived, who said; 'This man cannot cure them.' Then, addressing the knight, he added, 'Which do you prefer, to live with a single leg or to die with both legs?' 'I prefer', said the knight, 'to live with a single leg.' 'Then bring,' said the doctor, 'a strong knight with a sharp axe.' The doctor stretched the patient on a block of wood, and then said 'Cut off the leg with the axe, detach it with a single blow.' Under my eyes the knight gave a violent blow. He gave the unfortunate man a second blow, which caused the marrow to flow from the bone, and the patient died immediately. As for the woman, the doctor examined her and said, 'She is a woman with a devil in her head. Shave her hair.' They did so, and she began to eat again—like her compatriots—garlic and mustard. Her fever grew worse. The doctor then said, 'The devil has gone into her head.' Seizing the razor, he cut into the head in the form of a cross. Then he rubbed her head with salt. The woman expired immediately. After asking them if my services were still needed, and after receiving a negative answer, I returned, having learned from them medical matters of which I had previously been ignorant.

It seems reasonable to assume that orthodox medical treatment in ninth-century Ireland was no more sophisticated than that.

59
The Vikings

The year 824 is one of the few dates for which we have any documentary evidence in the history of Skellig Michael, but the events of Chapter Four are summed up in the *Annals of Inisfallen* very briefly. 'Scelec was plundered by the heathens and Etgal was carried off into captivity, and he died of hunger on their hands.' The heathens in this case were the Vikings, otherwise known to the Irish as Finngaill (the fair foreigners), whereas the later Norman invaders were known as the grey foreigners, possibly in reference to the chain mail they wore, which gave them a crucial advantage in combat over Irishmen who had no form of armour. The first Viking raid took place in 795, when Norwegian vessels came down the Scottish coast and crossed the water to attack the monastery of Rechru on Rathlin Island, County Antrim.

Typical of the craft which made these raids is the almost perfectly preserved vessel which was excavated at Gokstad, Vestfold, where it may have been constructed in the middle of the ninth century. It was 76½ ft long, 17½ ft in the beam, 6 ft 4 in from gunwale to keel amidships, clinker built, with keel and almost everything else in the hull made of oak; but its decking, made loosely for easy access to storage space underneath, was pine. Its sixteen strakes (nine underwater) were joined by iron rivets and small iron plates on the inside, and the hull had nineteen frames and crossbeams. Pine provided the mast and the sixteen pairs of oars, of different lengths so that they would strike the water together, after being put out through holes in the fourteenth strake: these could be closed when the oars were shipped again and the vessel was under sail. There was also a steering oar on the starboard side. The mast would have been anything between 26 and 35 feet high, and the boat's rectangular sail of heavy woollen cloth would have been about 23 × 36 ft in size. It is thought that the Gokstad ship, which weighed no more than twenty tons, had a crew of between thirty-two and thirty-five, and that few Viking dreki (dragon-heads, from the intimidating emblem carved on the bow) would have been bigger than that. The oars were generally used only in flat calms, or for a quick getaway after a raid: the Vikings at the time were by far the best seamen under sail in the waters that they cruised.

Within a few years these ships had extended their range down into the Irish Sea and were probing the Atlantic coast as well, and on both sides of Ireland they quickly discovered that the best loot was to be found in churches and other religious centres. From 850 there were also Danish incursions, but the greatest and most lasting impact was made by raiders who had sailed from the western seaboard of Norway, and who set a new course when they had passed the northwest tip of Scotland (Cape Wrath's name is derived from the Old Norse 'hvarf', which means 'point of turning'). We know this partly because Celtic ornamentation, often of ecclesiastical origin, has turned up in graves at Sogn og Fjordane, Hordaland, Rogaland, Vestfold and other sites on the west coast of Norway more copiously than anywhere else in Scandinavia.

The greatest raider of all was the fearsome Turgeis, who reached Ireland in 840 and thereafter became the principal bogeyman of all legends associated with the Vikings. He captured Armagh, and later sailed up the Shannon as far as Lough Ree, but is remembered best for the behaviour of his wife at Clonmacnois, where she is alleged to have performed heathen rites on the high altar. Nevertheless, it was Turgeis and men like him who profoundly altered the structure of Irish life for the first time and advanced its economy, by founding the first genuine urban communities in Dublin, Wicklow, Wexford, Waterford, Cork and Limerick. Not all the natives found them repugnant, some Irish abandoning Christianity to follow the Northmen and their gods, intermarrying and forming a mongrel sub-species known as the Gall-Gaedhil (foreign Irish) who might fight on either one side or the other, and were thus eventually regarded as uncertain allies by both Vikings and pure Celts.

There was at least one other raid on Skellig Michael, after the sally that cost Etgal his life. In *The War Of The Gaedhil With The Gaill,* an entry for the year 850 says that 'There came a fleet from Luimnech in the south of Erinn, they plundered Skellig Michael, and Inisfallen and Disert Donnain and Cluain Mor, and they killed Rudgaile, son of Selbach, the anchorite. It was he whom the angel set loose twice, and the foreigners bound him twice each time.' However many attacks were made on the skeilic in the ninth century, they would have perceptibly changed a monastic view of the world that the monks believed they had separated themselves from. After 824, never again could they feel secure and in control of their isolation, always they would uneasily anticipate another raid, forever the approaching stranger would be regarded with suspicion until he had demonstrated his benign intentions. It is therefore held to be quite possible that a monk of Skellig Michael was the author of an anonymous ninth-century verse which expresses a feeling that the raids would have engendered:

The wind is rough tonight,
Tossing the white-combed ocean.
I need not dread fierce Vikings,
Crossing the Irish sea.

The same sensations, of course, would have been experienced by all the Irish in the century following 795. For the first time in their recorded history they had been violated by the foreigner, who suddenly appeared in a role that they had regarded as their own prerogative before this. Subsequent incursions by Normans, English and Scots would only increase a sense of unease, suspicion, hostility and ultimate bitterness. It was the Vikings, though, who first drew the Irish into a world that everyone else was already all too familiar with.

63
Celtic Metalwork

The bell that the Vikings stole from the skeilic when they abducted Etgal would not itself have had more than intrinsic value, its worth depending entirely upon the beliefs of its owner. Irish bells of this period were wedge-shaped, in which some commentators have seen Eastern affinities, they were made from iron or bronze (sometimes a combination of the two) and they had a handgrip at the top so that they could be swung in order to produce sound: to that extent they were worth no more than the value of the metal. What made them much more precious to Christians was that they were often seen as holy vessels. The cult of relics had begun in Ireland from the moment Palladius preceded Patrick in 431, bringing with him, so it was believed, relics of the apostles Peter and Paul, which were subsequently kept in a casket at Killeen Cormac, County Kildare. After that, anything at all that might be associated with persons of great sanctity became the object of profound veneration, and were often used to ward off or cure disease, as in the case of the *Book of Durrow*, mentioned earlier (on page 152). Water drunk from any holy bell was generally thought to be especially efficacious in relieving boils or deafness. As late as the nineteenth century, jewels were prised from their settings on the casket containing St Patrick's Bell, not for their marketable value, but to be kept very carefully as charms against sickness or other misfortune.

Such cases (or shrines) were deliberately portable, because they and their contents were periodically taken on quite extensive progresses in order to impress and encourage the faithful over a wide area. They housed books and human remains as well as bells, and were themselves frequently of immense value beyond that of their special associations. Simply made of inexpensively sturdy materials at first, the shrines were later fashioned from precious metals with inset jewellery, and the craftsmen who made them were demonstrating a high talent for Celtic art which was at least equal to that of the monks who were creating the great illuminated manuscripts in their scriptoria. This form of Irish artistry reached its height in the eleventh and twelfth centuries, but it was sufficiently developed by the ninth century for the various kinds of portable shrine, together with communion vessels, book covers and other ecclesiastical objects made of precious metals with precious or semi-precious stones, to be prime targets of Viking raiders.

Because of this, very little of the most valuable early work has survived, and it is thought that one of the most marvellous examples of Irish metalwork is with us still (in the National Museum in Dublin) only because it was carefully hidden beneath a stone near the village of Ardagh, County Limerick, when the locals heard that the Vikings were on their way; there to remain until it was discovered by chance in the ruins of a ring fort in 1868. This Ardagh chalice is thought to have been made in the eighth century, and is basically a hammered silver bowl, in which the communion wine would have been served; but it has no fewer than three hundred separate components, with the bowl, the stem and the base being attached to each other with bronze pins. Gold, silver and copper wire has been used to add exceptionally delicate decoration in filigree, and the vessel is embossed in many places with enamelled studs, elsewhere with inset pieces of malachite, amber and rock crystal. The bowl has been lightly incised, partly with the names of the twelve Apostles; but the most striking decoration on the several parts of the chalice is very obviously kin to the knotwork and animal motifs that recur time and again on the pages of all the illuminated Celtic manuscripts. No civilisation has ever produced any objects more beautiful than the Ardagh chalice, the slightly later Derrynaflan chalice, and other such belongings.

The monks of Skellig Michael evidently hid their precious effigy of a kilted Christ crucified when one of the Viking raids threatened them. It was discovered in one of the oratories during the nineteenth century, by workmen who were building the first lighthouse on the skeilic. A Mr T. Crofton Croker made a drawing of it at the time, which was fortunate, because it disappeared again shortly afterwards and has not been seen since.

75
Skellig Birdlife

The proximity to and awareness of nature that characterised so much Celtic art was nowhere more intense than it would have been for the monks of Skellig Michael. There is no evidence whatsoever that any form of that art ever originated on the skeilic, no reason to suppose that any illuminated text was composed there and was subsequently lost. But if the community had run to a scriptorium, we may be sure that the pages coming out of it would have been well decorated with marginal fish and bird images at least. Of mammals there would probably have been nothing at all other than seals, though Skellig Michael today also hosts both rabbits and mice. The birdlife, however, would have been almost as vibrant and prolific as it is now. Though it is impossible to say with certainty which species made their home on the skeilic or on the adjacent Skellig Beag in the tenth century, it seems likely from what is known about the region's marine birdlife in general at that time that the gannet (gainéad), the puffin (puifín), the razorbill (crosán), the guillemot (forach), the kittiwake (staidhséar), the storm petrel (mairtíneach), the cormorant (broigheall) and the shearwater (cánóg), as well as different varieties of gull (faolieán) would all have been familiar to the monks as part of their own habitat; there may also have been isolated examples of the raven (fiach), the chough (cág cosdearg), the falcon (fabhcun) and the wren (dreoilín). All these birds are to be seen on or from Skellig Michael nowadays, and many of them nest there, but one of the commonest residents now was certainly missing then. The fulmar, for which there is no name in Irish, was unknown anywhere in Ireland or the British Isles except on St Kilda before 1878, and it was first reported on Skellig Michael only in 1913. It is now there in great quantity from January to September every year.

80
The Dancing Sun

The phenomenon which Ciarán beheld at sunrise on Easter morning lies at the very heart of the naturism embraced by Celtic Christianity, as does his interpretation of it. It is said that even within living memory many Irish habitually climbed hills on Easter morning in the hope of seeing the new sun acknowledge the risen Christ by some unusual movement; and the same instinct has been recorded along the western seaboard of Scotland. An account of the phenomenon appears in the *Carmina Gadelica,* where Alexander Carmichael reported that:

The people say the sun dances on this day in joy for a risen Saviour. Old Barbara Macphie at Dreimsdale saw this once, but only once, in her long life. And the good woman, of high natural intelligence, described in poetic language and with religious fervour what she saw or believed she saw from the summit of Benmore: 'The glorious gold-bright sun was after rising on the crests of the great hills, and it was changing color—green, purple, red, blood-red, white, intense white, and gold-white, like the glory of God of the elements to the children of men. It was dancing up and down in exultation

at the joyous resurrection of the beloved Saviour of victory.' To be thus priv-
ileged, a person must ascend to the top of the highest hill before sunrise, and
believe that the God who makes the small blade of grass to grow is the same
God who makes the large, massive sun to move.

It has been remarked that the most significant thing about that
anecdote, given the enormous assumption behind it, is that Barbara
Macphie claimed to have seen the sun dancing only once in a long life.
Had she gone looking for it, had she wanted it to happen, had she
been intent above all on impressing others with something manifested
to her alone, the wish might well have become father to the thought,
and she would doubtless have told of recurrences. The singularity of
her sighting makes an extraordinary movement of the sun on Easter
morning all the more credible. The interpretation of that movement
is, of course, another matter. But it has been pointed out that Barbara
Macphie's experience, whatever the reality of the experience, may
have been touched by the same mystical perception as something
recorded by Patrick in his *Confessio.* The saint was troubled in a dream
by Satan, who 'fell over me like a huge rock, and none of my members having any prevailing power . . . And amidst these things I saw
the sun rise into the heaven, and while I was calling "Elia, Elia" with
all my powers, look, the splendour of His sun fell down over me, and
immediately shook off from me all oppressiveness, and I believe that I
was come to the aid of by Christ my Lord . . .' In both cases, it has been
argued, the sun is not a metaphor or an image of Christ, but a medium
through which Christ shines.

85
The Irish Annals

In the *Annals of Inisfallen,* it is recorded that in 1023 there was 'Great drought from the Epiphany until May . . . A solar eclipse this year, in the spring of the black cloud.' And for the year 1044 the *Annals* note that 'Aed Sceilic, the noble priest, the celibate, and the chief of the Gaedil in piety, rested in Christ.' These *Annals* are associated with the monastery which was founded by Faithliu, son of the king of West Munster, in the seventh century on the island of Inisfallen, which stands in the middle of Lough Leane, near Killarney. They are the principal existing record of Munster's medieval history, from 433 to 1326, but only the later part of that chronology was written on Inisfallen, the earlier work almost certainly being composed about 1092 from documents belonging to the monastery at Emly, County Tipperary. Many similar chronicles were made elsewhere, covering other areas of Ireland and different periods, among them two different *Annals of Boyle,* one of which begins with the creation and runs to 1253, the other from 1224 to 1562; the *Annals of Ulster,* the *Annals of Clonmacnois,* the *Annals of Tigernach,* the *Annals of Connacht,* the *Annals of Loch Cé.*

The greatest compilation of all is generally agreed to be the *Annals of the Four Masters,* which was chiefly the work of the Franciscan Brother Michael O'Clery, who was born in Donegal in 1580. O'Clery had been something of an historian or antiquary before his profession as friar, and after being sent to work at the Irish College in Louvain, he obtained permission to scour the monasteries of his native land for their manuscripts and vellum books, of which he sent copies back to the library in Flanders. He started this work at a convent near his birthplace in 1632 and finished it four years later, with the assistance of three others (hence the Four Masters). Two of these were kinsmen, including Conary O'Clery, who acted as a scribe for his brother Cucogry O'Clery, a dispossessed landowner and head of the Tirconnell sept of the O'Clerys. The fourth master was Ferfeasa O'Mulcrony, of whom we know nothing except that he was an antiquary. Of their work, which covers Irish history 'from the earliest period to the year 1616' in six volumes, Douglas Hyde once wrote that 'So long as Irish history exists, the *Annals of the Four Masters* will be read.' But in the entire work, Skellig Michael is mentioned only three times. 'In the Age of Christ 823 [sic] Etgal of Scelig was carried off by Gentiles, and died soon after of hunger and thirst . . . 950 Blathmhac of Sgeillac died . . . 1044 Aedh of Sgelic-Mhichíl, and Ailil, son of Breasal, resident priest of Cluain-mic-Nois, died . . .'

86
Brian Boru

There is reason to believe that Inisfallen was the place where Brian
Bóruma (Brian Boru) was educated, before becoming ard rí of
all Ireland and posthumous victor at the Battle of Clontarf in 1014.
This has entered local mythology as the occasion when the Irish king
led his people to defeat the Vikings gloriously at the cost of his own
life, but the reality of Clontarf was a bit more complicated than that.
For one thing, Brian's opponents were a coalition of local Norsemen,
together with men from Orkney, Shetland and Man, as well as native
Irish from Leinster who resented the Munsterman's high kingship,
which had ended centuries of supremacy by chieftains of the Ui Niall.
For another, the Norse ruler of Dublin, Sihtric, who was Brian's
brother-in-law, took no part in the battle just outside his town, while
the Limerick Vikings sided with Brian for jealous motives of their
own. There were similar and all too familiar feuds which set the Irish
of Connacht as well as the Leinstermen against Brian, who had attacked his predecessor Malachy by sending Norse vessels up the Shannon to ravage the old ard rí's lands. Nor did Clontarf result in the
expulsion of the Norse from the Irish towns they had created. They

were still there a century later, having a generally beneficial effect on urban life and national commerce, while the Irish kings resumed their internecine struggles until some sort of unity was forged in hostility to the Normans and English. The only clearcut thing about the Battle of Clontarf was its dreadful casualities, with 4,000 on Brian's side killed and perhaps 7,000 of their opponents. Brian was not the only member of his family to die in the battle, cut down as he prayed for victory in a grove of trees: his son was killed shortly afterwards, when victory was assured, and his grandson drowned near the Weir of Clontarf when he was hunting enemies who had fled the field.

86
Olaf Tryggvason

The Norse king who sought Christian baptism was the Olaf Tryggvason whose youth and early manhood were a paradigm of Viking adventure. Born *c.* 968 after his mother Astrid had fled from a massacre at Oslofjord, in which his father Tryggvi Olafsson was slain, Olaf's first three years were fugitive in eastern Norway and Sweden. He survived capture by Estonian pirates, was ransomed from slavery by an uncle who had high rank at the Russian court in Novgorod, and at the age of eighteen began his own viking (roving, raiding) career in the Baltic. By the early 990s he was raiding the British Isles as an ally of the Danish king Svein Forkbeard and, according to his thirteenth-century biographer Snorri Sturlusson, he ranged from Northumbria and Scotland, through the Hebrides, Man, Ireland, Wales and Cumbria till he sailed to France. It is accepted that Olaf became Christian before 995; but there has always been uncertainty about where the conversion took place.

Snorri mentions a visit by Olaf to the island of Skyllingarna, where a hermit prophesied a rosy future. Some sources have identified this place as the Scilly Isles, but an attempt has also been made to associate

Olaf with Skellig Michael instead, without any other evidence, even going so far as to suggest that it was on the skeilic that Olaf was baptised. The best authorities are inclined to put their trust in *The Anglo-Saxon Chronicle,* where it is recorded that in 994

Olaf and Swein came to London, on the Nativity of St Mary, with ninety-four ships, fighting constantly the city, and they meant, moreover, to set it on fire. But they suffered more harm and evil than they ever believed any town-dwellers could have done them. In this God's holy mother showed her mercy to the town-dwellers and delivered them from their enemies; then they went from there, and wrought the most evil that any force had ever done, in burning, ravaging and killing, both along the sea-coast, in Essex and in Kent, Sussex and Hampshire; finally they seized horses for themselves, and rode as widely as they would, working unspeakable evil. Then the king and his counsellors advised that they be sent to, and tribute and provisions promised them, so that they would leave off harrying. Then they accepted that, and the whole force came to Southampton, and took winter quarters there. They were provisioned throughout all the West Saxon kingdom, and they were given sixteen thousand pounds. Then the king sent bishop Aelfheah and ealdorman Aethelweard to king Olaf, and meanwhile hostages were exchanged in the ships. Then they led Olaf, with much celebration, to the king at Andover. King Aethelred received him of the bishop's hands, and gave him kingly gifts. Olaf then promised him—and also did as he promised—that he would never again come to the English people in enmity.

The king's reception of Olaf 'of the bishop's hands' does seem to indicate confirmation of baptism at Andover in 994; and it is probably safe to assume that the baptism immediately preceded it (it is unlikely that an already baptised Olaf would have wrought the earlier havoc among Saxon co-religionists). Moreover, Olaf sailed for home shortly afterwards, taking with him priests who had been trained in England. He was proclaimed king of his ancestral lands at Trondheim in 995 and began to spread the faith, though there is argument about the depths of his spiritual beliefs and whether they convinced him more than the political advantages he saw in attaching Norway to the Christian community of Europe. There is also doubt about the circumstances of his death, which occurred five years later, when he went missing, and was presumably drowned, during a sea battle whose details vary from one historical source to another.

86

The Culdees

The Culdees were another of those familiar landmarks in the history of the religious life, the reforming movement which gathers momentum because it is felt that the purity of an original doctrine has been defiled, or that those under vows have simply become lax in their observances (thus the Cistercians were a reaction to the perceived failings of Benedictines, and the Trappists subsequently took the view that the Cistercians themselves were no longer rigorous enough). The word Culdee is a corruption of Céli Dé, which translates as Servant of God, though this was frequently rendered more loosely as Companion of God; and the Culdees often referred to themselves as the Sons of Life or Light. Insofar as the movement was inspired by an individual, he is the rather forbidding figure of Mael Ruain, who founded the monastery of Tamlachta (Tallaght, Dublin) in 769, from which other houses issued later at Finglas, on the other side of the Liffey, and at Terryglas on the Shannon. One of his less attractive characteristics was codified in the extensive *Rule of the Céli Dé;* his instruction that 'during the monthly disease which is upon virgins of the Church . . . they

shall not go to hand then, because they are unclean during that time.'
Going to hand meant receiving the priest's blessing after confession.

Mael Ruain was an obsessive ascetic and, as a result, some of the
most severe dietary restrictions in the history of Irish monasticism
were practised by his followers, as when a community of Culdees
spent an entire year on bread and water alone, so that the soul of a recently dead brother might be more surely released. Mael Ruain was a
great believer in psalms and other prayers being said crosfigel, and in
plenty of them; also in lengthy penances being exacted for failure to
meet the norms. The discipline (that is, self-flagellation with knotted
cords) was regularly applied by Culdees, though scourging was as yet
relatively infrequent at this stage of Irish Church history. It was also
the Culdees who notoriously put themselves helplessly in the hands of
Providence by casting themselves adrift upon the waters, accepting
whatever the outcome might be. But pilgrimage was discouraged,
and on no account was a Culdee to ask for any news of the outside
world if his house received visitors.

And yet from this extremely austere regime emerged some of the
most tender nature verse that the Celtic imagination ever produced,
and anchorites of the Céli Dé became celebrated for their ability to
coax wild creatures into companionship. It was such a one, very likely,
who in the ninth century wrote the 'Hermit's Song', which begins:

I wish, O Son of the living God, O ancient, eternal King,
For a hidden little hut in the wilderness, that it may be my dwelling.

An all-grey lithe little lark to be by its side,
A clear pool to wash away sins through the grace of the Holy Spirit.

Quite near, a beautiful wood around it on every side,
To nurse many-voiced birds, hiding it with its shelter.

A southern aspect for warmth, a little brook across its floor,
A choice land with many gracious gifts, such as be good for every plant ...

Although there would also be Céli Dé in several parts of Scotland and, further south, at York and Bardsey Island, the movement's energy eventually died, and the Culdees sank back into the generality of Church life, the position they had developed within its structures superseded to a large extent by the arrival of canons regular, notably those following the *Rule of St Augustine*. They were turned out of Down Cathedral in 1183 by the Norman John de Courcy, who brought Benedictine monks from Chester, rather than Augustinians, to form the chapter in their place. But by then, the whole thrust of Irish Christianity had swung from a monastic to a diocesan emphasis.

94
Giraldus Cambrensis

We can only surmise that the dedication of the skeilic to the Archangel Michael took place just before the middle of the eleventh century, because the evidence is imprecise, and effectively limited to two of the entries already cited from the *Annals of the Four Masters*. In it there is a reference 'in the Age of Christ 950' to one Blathmac of Sgeillac; whereas in 1044 Sgelic-Mhichíl is mentioned for the very first time, in connection with the death of Aedh. We are on similarly uncertain ground about the date when the chapel was built, the architectural evidence indicating merely that this could not have been earlier than the eleventh century. Beyond that, we have no one but Giraldus Cambrensis (Gerald of Wales) to turn to. Three parts Norman in fact, and only one part Welsh, he was born in Pembrokeshire *c.* 1146 and never realised his lifelong dream of becoming Archbishop of Wales, having to be content with a subordinate role as Archdeacon of Brecon instead. But he was a great man for knowing people who mattered (he would be called a terrific networker today), and he wrote racily about the affairs of his time in seventeen books. One of these was his *Topographia Hibernica et expugnatio Hibernica*

(The History and Topography of Ireland) which he composed after first visiting the country in 1183, then again in 1185. Without it, we should know a great deal less than we do about the state of Ireland in the twelfth century, but it suffers from a number of very obvious limitations. One of these is the fact that Gerald never saw the Atlantic seaboard of Ireland: he may have reached the Shannon near Athlone, but he certainly did not travel any further west. So anything that he reports about County Kerry or any of its offshore islands can be nothing more than hearsay, including the following apparent reference to Skellig Michael:

In the south of Munster near Cork there is a certain island which has within it a church of Saint Michael, revered for its true holiness from ancient times. There is a certain stone there outside of, but almost touching, the door of the church on the right-hand side. In a hollow of the upper part of this stone there is found every morning through the merits of the saints of the place as much wine as is necessary for the celebration of as many Masses as there are priests to say Mass on that day there.

96
The Hermitage

We are similarly in the dark about the date of the hermitage just below the summit of Skellig Michael's so-called South Peak, which ought more accurately to be identified as the West Peak. All that can be said with complete confidence is that it was constructed sometime between the eighth and thirteenth centuries, the likelihood being that the work was not done at the end of that timespan. The dating is, in truth, almost irrelevant to the most important thing about the hermitage, which is the very fact of its existence on such a hair-raisingly precarious site. To reach its three terraces just below the summit of the peak calls for nothing less than recognised rock-climbing skills, in order to negotiate the chimney known as the Needle's Eye, to make traverses along narrow ledges above a long drop into the sea, and to take advantage of similarly exposed foot- and handholds that the monks found or put in the almost vertical cliff face. The complexity and thoroughness of the construction up there, which resulted in an oratory, an altar, a cistern and a minimal form of shelter for the incumbent, as well as several retaining drystone walls, can only be a matter for astonishment. There was also a small patch of ground with light and dry soil which would have drained well, and on which a few vegetables could have been grown. The entire enterprise has been summarised in the following terms:

a hermitage built virtually in the air on the treacherous ledges of an Atlantic rock rising straight up from the ocean to an altitude of 218 meters [715 ft]... On the way up this traverse one views with amazement a fragment of drystone wall built by someone who must have been kneeling on clouds when he placed those stones on a narrow ledge that plummets into what appears eternity.

The hermitage could only have been the product of almost super-human determination and zeal by one individual, aided and abetted by a group of equally dedicated co-workers in the construction. It has been pointed out that anyone who took it upon himself to live in soli-tude in such spectacularly hazardous circumstances, facing the possi-bility that he might well be swept to his death in the first winter storm that smashed into the skeilic, was not only fortified by remarkable tenacity of purpose and a profound faith, but was perhaps suffering from a rather dangerous form of spiritual arrogance, and was in a very real sense indifferent to death. 'The first reaction is to gasp at the dar-ing construction,' writes Professor Horn,

to marvel at the stunning Atlantic beauty of the site and to wonder, Why? Why was this ever built, this wall hanging on the edge of space? The first her-mit who left the monastery was going in one direction only—up, closer to God through a life of complete isolation on the island's highest point... As one contemplates this construction, one wonders whether at some point devo-tional submissiveness might have shifted imperceptibly into devotional hubris.

The most obvious candidate for such a role would have been one of the Culdees; but linking such a candidature with the established date for the death of Aedh can be no more than another speculation. Set beside the breathtaking reality of the hermitage's existence at some point in the given timespan, however, the true identity of the hermit is almost neither here nor there.

100

Origen

A clinical diagnosis of Aedh's erotic and other visions would doubtless have concluded that, whatever shaped them in his psyche, they were triggered by his reckless fasting. Hallucination as a result of extreme exhaustion, including that which has resulted from semi-starvation, is a well-established condition, though many more centuries would pass after Aedh's time before this was recognised. But visionaries of every faith at all stages of history have tended to be people whose lives are marked by exceptional austerity, and it is difficult to think of a single instance in which a holy man or woman has reported tempting, fearsome or inspiring manifestations, on a regime of ample food and sleep. Hallucinations are also well known in the case of many prisoners held captive in the most appalling conditions.

The Origen who crossed Aedh's mind was certainly one of the third century's great ascetics, but he did not become celebrated as a pictorial visionary: his fame spread because of his cerebration and powers of analysis that combined Hellenistic thought with biblical scholarship. He believed that the interpretation of Scripture was his vocation, and saw the historical narrative as secondary to its disclosure of spiritual truth. He drew much of his intellectual vigour from Plato, yet remained sure that the Bible was the only source of divine revelation. But the fact that he could even discuss rationally whether there was such a thing as 'freedom in God' or whether God was simply an all-powerful creator of the cosmos, caused Origen to be regarded as heretical in some quarters, especially with Bishop Demetrius of Alexandria, who had given him his first teaching promotion. He was still out of favour when he died in Palestine *c.* 254, but his teachings were held in high esteem long afterwards by, among others, John Cassian.

Origen was born at Alexandria in 184, into a Christian family, and at the age of eighteen, when the persecutions of Septimius Severus began, he urged his father to accept martyrdom without worrying about the abandoned wife and children; would, indeed, have joined his father if his mother had not hidden his clothes. His subsequent asceticism was doubtless—as in the case of Irishmen later—a substitute for the real thing. So may have been the notorious event which Aedh recalled on the terrace of his hermitage. It has always been understood (though this is contested by some, who think it a malicious piece of gossip retailed by his biographer, Eusebius of Caesarea, from hearsay) that Origen castrated himself after taking too literally the injunction in Matthew 19:12 ('there are some who have made themselves eunuchs for the kingdom of heaven's sake'), which Antony the Hermit a generation later would cite as one of the two chief authorities in support of the monastic life. Indisputably, however, there were such examples of self-mutilation in the early Church.

101

The Bestiaries

The usually alarming and utterly strange creatures that were manifested to Aedh in his exhausted stupor were familiar figures in the medieval imagination. They had infiltrated from a variety of sources, starting with a medley of Greek and Roman authors, who were either natural philosophers or storytellers peddling the heroic legends. Herodotus, for example, had put about the myth of the phoenix which, every five hundred years, visited the city of Heliopolis with the body of its father embalmed in myrrh, and interred this reverently before plunging to its own death in fire and then being reborn from the ashes; Pliny, whose *Historia Naturalis* ran to thirty-seven volumes, was responsible for awareness of the Sciapod, which moved with astonishing speed in spite of its single ten-toed leg, this also serving to shade it from the sun; and the Homeric epics are well populated with fearsome apparitions, of whom Scylla and Charybdis are only two. Travellers' tales of things seen in distant lands also played their part in feeding the medieval mind with intimidating images, and people were perfectly capable of constructing impossible creatures out of tittle-tattle from the next parish. Giraldus Cambrensis, who obviously did not see any such thing, mentions the Irish hybrid that 'had all the parts of the human body except the extremities, which were those of an ox . . . He could not speak at all, he could only low . . . He came to dinner every night and, using his cleft hooves as hands, placed in his mouth whatever was given him to eat . . .'

An important key to this phantasmagoria was the text known as the *Physiologus,* which was written in Greek, possibly by an Alexandrian contemporary of Origen, who may also have been another ascetic. While appearing to be a straightforward study of animal behaviour, with a few dozen examples, its real purpose was to draw conclusions from that behaviour which would tally with and reinforce the Christian philosophy. It noted the phoenix, for example, which sacrificed itself and was then restored to life, thus illustrating the truth of Christ's resurrection. The *Physiologus* was to be translated into many languages, and its Latin version in particular, which was available by the fifth century, was especially influential. Among those fascinated by its contents was the seventh-century Bishop Isidore of Seville, whose own greatest celebrity would be the publication of an *Etymologiae* in twenty volumes, an encyclopaedia of many topics, with volumes XI and XII covering men, fabulous monsters and animals.

From such beginnings the medieval bestiaries developed, hand-copied and well-illustrated manuals which served both to entertain and instruct a credulous society. They were, of course, produced in the monastic scriptoria and no monastic library would be regarded as complete without as many bestiaries as the scribes could lay their hands on and reproduce. They were suggestive reference books for homilies on the standards that were expected of all Christians, on the perils that awaited them if they fell from grace; also to illustrate the infinite variety and wondrous ingenuity of God's creation. Novice monks would be instructed in these matters, and so would the secular congregations that thronged the monastic churches. The bestiaries also influenced the wider audience when stone masons consulted them at the building of churches, before carving their grotesque rainwater heads and other gargoyles. And they left their mark on cartography, by providing illustrations for the more outlandish and uttermost parts of the earth. Neither the fourteenth-century *Mappa Mundi,* which hangs in Hereford Cathedral, nor the thirteenth-century Ebstorf map in Germany (which was tragically lost in bombing and can now only be enjoyed in facsimile) would be half as rewarding as they are, were it not for their copious inclusion of these mythical cartoons.

So the spooky creatures which were paraded before Aedh would have had a wide currency in the subconsciousness of eleventh-century Ireland. Apart from the Indian Sciapod, these included the African Blemyes, or Anthropophagi, the men 'whose heads do grow beneath their shoulders'; the bat-like Phanesii from Asia, who 'neede none other apparell to clothe theyr limbes with, than thyr own flappes'; the dog-headed Cynocephales, who originated in Norse mythology; the slithering four-legged Himantopodes from Africa; the Manticora from India, part-man, part-lion, part-scorpion, which revelled in human flesh, was swifter than a bird, and had three stings which could be used as darts; the Asian Yale, with its horse's body, goat's jaws and straight horns, which centuries later turned out to be not mythical at all, but simply a rather misleading description of the Indian water buffalo; African ants, which guarded golden sands and threatened anyone who looked likely to purloin any; and the Phrygian Bonnocon, which discharged its fiery dung over an area of three acres when chased—he appears on the *Mappa Mundi* and has a particularly smug expression on his face.

104
Hubris

Aedh's terrible death was not without precedent in the history of the ascetic life. In his second Conference, Cassian tells a story to warn all potential anchorites about the dangers of getting things out of proportion. It concerns the old hermit Hero, who spent half a century in the desert and pursued his Christian ideal with a rigorous fervour that everyone marvelled at, because it so outdistanced anybody else's attempts at the eremitical life. But disaster awaited him, as Cassian relates.

He practised fasting so rigorously and so relentlessly, he was so given to the loneliness and secrecy of his cell, that even the special respect due to Easter day could not persuade him to join the brethren in their meal. He was the only one who would not come together with all his brethren assembled in church for the feast, and the reason for this was that by taking the tiniest share of vegetables he might give the impression of having relaxed what he had chosen to do.

This presumptuousness led to his being fooled. He showed the utmost veneration for the angel of Satan, welcoming him as if he were actually an angel of light. Yielding totally to his bondage, he threw himself headlong into a well, whose depths no eye could penetrate. He did so trusting completely in the assurance of the angel who had guaranteed that on account of the merit of his virtues and of his works he could never come to any harm. To experience his undoubted freedom from danger the deluded man threw himself in the darkness of night into this well. He would know at first hand the great merit of his own virtue when he emerged unscathed. He was pulled out half-dead by his brothers, who had to struggle very hard at it. He would die two days later. Worse, he was to cling firmly to his illusion, and the very experience of dying could not persuade him that he had been the sport of devilish skill. Those who pitied him his leaving had the greatest difficulty in obtaining the agreement of abbot Paphnutius that for the sake of the merit won by his very hard work and by the many years endured by him in the desert, he should not be classed among the suicides and, hence, be deemed unworthy of the remembrance and prayers offered for the dead.

105
The Withdrawal

The withdrawal of the community from Skellig Michael again involves a certain amount of speculation, though there is a great deal of circumstantial evidence which leads to a number of conclusions. Judging from entries in the *Annals of Inisfallen,* the weather does seem to have deteriorated significantly in the last quarter of the twelfth century, though the worst of it would not arrive until the thirteenth century was well under way, after which the *Annals* reported terrible storms and snowfalls (also an earthquake) on an almost annual basis: including '1222. A great wind throughout Ireland, and it wrecked houses, churches, and great woods, and sank many ships, and it was known as the great wind . . .' But given that monks had by then inhabited the skeilic for several hundred years, in which severe storms were a feature of every winter and were liable to blow up at any time, including midsummer, it seems improbable that even a sustained climatic change for the worse would alone be enough to drive them from what had become their natural habitat. There must have been other factors which, coming on top of deteriorating weather, made up their minds that they should start afresh on the mainland.

It has to be remembered that thirteenth-century Ireland was not the same country that produced Celtic zealots seeking a white martyrdom offshore in the sixth century. A social transformation had begun with the coming of the Vikings, and it had been accelerated from the moment the Normans appeared and began to colonise in a manner that had never been the way of the Norsemen. They came with the blessing of Henry II but at the invitation of Dermot Mac-Murrough, king of Leinster, who was losing a power struggle for the high kingship with Rory O'Connor of Connacht; and when Dermot died in 1171 the victorious Richard de Clare (Strongbow), whose mail-clad troops had carried all before them, and who was by now married to Dermot's daughter, became ruler of Leinster in his stead. For the first time King Henry personally intervened in Ireland, not so much to conquer it as to make sure that the *arriviste* Norman lordlings there did not become powerful enough to threaten him. Conquest, nevertheless, is effectively what followed, though many sections of Irish society, including most of the native rulers, welcomed the one man who might curb local Norman excesses.

The leaders of the Irish Church, too, were complaisant. The Church had, as we know, been changing course from its Patrician origins since just before the Synod of Whitby, and it had been profoundly marked by the depredations of the Norsemen, which irreparably damaged the settled and dominant life of the monasteries. Leadership thereafter began to shift from abbots to bishops, and other changes were set in train, but it was not until the arrival of the Normans that the Irish learned how much their isolation was now perhaps no longer the blessing it had always seemed, but a disadvantage if their Church was to thrive as of old. Already, before the middle of the twelfth century, a continental style of territorial dioceses had been established, with archbishops in Armagh, Dublin, Cashel and Tuam who had each received the pallium (a band of white woollen material with several purple crosses, worn hanging from the shoulders and loaded with Catholic significance) personally from the Pope, in acknowledgement of Rome's absolute authority, and its delegation to the appointed archbishop.

One of the principal native reformers was Mael Maedóc O Morgair, who was formally canonised as St Malachy within fifty years of his death in 1148. Born in Armagh at the end of the eleventh century, he became bishop of Down at a time when Irish Christianity had become lax and without a sense of mission, from which he was determined to restore it to its former vigour and piety. He was nominated to the archbishopric of Armagh, but was bitterly opposed by the family which regarded that see as its hereditary possession. So in 1139 he journeyed to Rome to seek the pallium from Innocent II, who refused it, but instead made Malachy his papal legate for Ireland. And for eight years, Malachy laboured hard at home before setting out once more for Rome to ask another Pope, Eugene III, for the archiepiscopal authority that Innocent had denied him. This time, Malachy never reached the Tiber: he died on the way, at the Cistercian abbey of Clairvaux in Burgundy, which he already knew well from his earlier journey. On his way to Rome in 1139, in fact, he had left four Irish monks to be trained there under St Bernard, and these men returned to Ireland with others in 1142 to found a Cistercian monastery at Mellifont, County Louth (they thus preceded the more senior Benedictine order to Ireland by almost half a century and, even today, Cistercian houses are much more numerous there). Malachy was also instrumental in introducing the canons regular to his country, consecrating an Augustinian house at Knock, County Clare, just before leaving Ireland forever.

All these changes effectively spelt the end of the traditional Irish monastic system, and replaced it with continental models of the coenobitic life. For one thing, the old abbatial autonomy had been superseded by submission to a diocesan authority who might not be any kind of monk at all. For another, the Cistercians, the Benedictines, the Augustinians and any other species of religious who now settled in

Ireland but were still in obedience to the mother houses of their orders in continental Europe, followed a Rule of conduct which had been drawn up by each founder and which was designed to produce a strict uniformity in everything—from liturgy to victuals, from the division of time between worship and labour to the special duties of the cellarer—in every single house of that foundation. Apart from the *Rule of the Céli Dé,* such codified discipline had been unknown in Ireland earlier, because each Celtic monastery was, in practice as well as in theory, a law unto itself, even though it might gladly acknowledge its pedigree and follow its parent in all essentials; as, for example, the monastery of Kells long remained basically faithful to the precepts of Iona. The point was, though, that the old-style Celtic abbots felt free to stamp a community with their own individuality, and most of them did so without hesitation. This latitude was no longer available by the end of the twelfth century.

There was one other reason which might well have impelled the monks of Skellig Michael to enter the new mainstream of Irish religious life, by leaving their skeilic for the mainland. Over the centuries its fame as an anchor-hold of exceptionally holy men had spread throughout Ireland and even beyond; and it is in the nature of all such places to attract the pilgrim and the merely curious, often in quantity and invariably despite difficulties in reaching the goal. We simply do not know when pilgrimage to Skellig Michael began, but the building of the 'new' chapel (whenever that was) is very probably a clue, together with the dedication to St Michael. There would have been no reason for a larger place of worship unless it was to accommodate more worshippers than had been the norm for hundreds of years; and we do know (or think we know) that the resident population of monks varied little across the centuries. An appreciable and regular influx of visitors would make the contemplative life, as generations of monks on the skeilic had understood it, very difficult to sustain in such confined circumstances. Better by far, the community might well

have concluded, to move out and adapt themselves elsewhere, switch-
ing the focus of everyone else's attention away from the skeilic, so that
it could become a place to which individual monks alone might retreat at intervals for periods of uninterrupted solitude; and thus regain
its original purpose.

106
St Benedict's *Rule*

The *Rule of St Benedict,* which sounded impossibly lenient to Abbot Eoghan, certainly took a view of the monastic life that was markedly different from the Irish tradition. Benedict, whose life in Italy was contemporary with that of Finnian of Clonard, was concerned to produce order out of a system which so far had tolerated the individuality of abbots everywhere, in spite of earlier attempts to regulate that system, first by St Pachomius in the Egyptian desert, then by St Basil the Great in Cappadocia. Irregularity had too often become chaotic, and Benedict thought to obviate chaos through legislation that would leave not the slightest opportunity for individual initiative. In that sense, the seventy-three chapters of his *Rule* represented an unyielding code of conduct. But the temper of the document is one of humane mildness throughout, and the amounts of food and sleep enjoyed by Benedictine monks, which would have astonished any Irish counterpart by their generosity, are merely one example of this. The Benedictine was also to be provided with adequate clothing, so that he would never be cold in winter, or stifled in the heat of an Italian summer; if a monk erred in some way, instead of being given some crushing penance at the outset, he should first be comforted by some older and more prudent brother 'lest he be swallowed up with overmuch sorrow'; and the abbot was himself reminded that he had undertaken 'the charge of weakly souls, and not a tyranny over the strong'. The whole tenor of Benedict's *Rule* is set down in one sentence of its prologue: 'Therefore must we establish a school of the Lord's service; in founding which we hope to ordain nothing that is harsh or burdensome.'

It has been plausibly suggested that one reason why the Benedictine *Rule* became the dominant code throughout western monasticism is that its very mildness made it much the most appealing: it did not promote a soft life by any means, but its basis was not perpetual hunger and corporal punishments of one sort and another. Another reason for its popularity might be that the *Rule* comprehensively covered every aspect of monastic life, and therefore ruled out demoralising uncertainty. Lastly, it received the backing of the Holy See, from the time of Gregory the Great, and as long as the Western Church in general remained obedient to Rome, this counted for much. But the fact that at Cîteaux in 1098 a reforming order was founded to recover a strictness that the Benedictines appeared to have lost was an indication that for some men and women, mildness would always seem the antithesis of the monastic life. Within twenty years, the austere Bernard of Clairvaux had joined the Cistercians and founded a daughter house, and it was perhaps no coincidence that the Irishman Malachy immediately befriended him, rather than any one of the many Benedictine abbots he must have come across on his journeys to and from Rome.

115
Ballinskelligs and the Arroasians

The place where Eoghan and his monks landed after leaving their skeilic was on the northern shore of Ballinskelligs Bay, and the penultimate conjecture about Skellig Michael is the date when this migration took place. The ruin that stands above the foreshore today is what's left of Ballinskelligs Abbey, and most of the remaining walls were built no earlier than the fourteenth century. There is, however, on the east end of the Prior's Lodge, a window which has been identified as *c.* 1270 and, close by it, a doorway in a style that belongs to the end of the twelfth century: in other words, most of the abbey was rebuilt, quite possibly more than once, and almost nothing of the first structure has survived. We can only be sure that the original foundation was either late in the twelfth or early in the thirteenth century.

The abbey belonged to the most important branch of the canons regular of St Augustine, the Arroasian Order, which combined the contemplative ideal in monasticism with the parochial work that played an increasing part in the life of religious under the new Irish diocesan emphasis. The order had Irish connections from the outset, when two chaplains to William the Conqueror, Heldemarus and

Conon, retired to a hermitage in 1090 near the town of Arrouaise in
Flanders, which was sited on the spot where two Irish saints, Luglius
and Luglianus, had been murdered by robbers several centuries earlier. Heldemarus, too, was killed within a few years, but Conon led
the new community to prosperity under the patronage of Lambert,
Bishop of Arras, who consecrated the first abbot of Arrouaise, Ger-
vase, in 1121. By 1148 there were daughter houses well established
elsewhere in Flanders, as well as in France, Germany, Poland and England; and that year saw the first Irish foundation. St Malachy had visited Arrouaise during his first journey to Rome and it is possible that
he left some Irish monks to be trained there, just as he certainly left
others to be tutored in Clairvaux. By the time the Arroasians settled
in Ireland, they were one of the most celebrated congregations of Augustinian canons, and they soon spread themselves widely among the
Irish: Christ Church Cathedral, Dublin, became an Arroasian holding in 1163.

What we do not know is how, or precisely when, the monks from
Skellig Michael made their connection with the Augustinians. The
most likely sequence is that almost as soon as the monks arrived from
the skeilic, they were in touch with the old abbey of Inisfallen, which
had long been their nearest neighbour, and which had adopted the
Arroasian Rule sometime in the second half of the twelfth century. It
would then have been logical for them to join forces under the new
dispensation, and for a daughter house of Inisfallen to be raised on the
shore of Ballinskelligs Bay. That a connection with Skellig Michael
was retained afterwards is beyond doubt, because at last we are into a
period of history where systematic records were kept. And in the taxation lists for the Ardfert diocese covering the years 1302–6, the
church of St Michael's Rock was taxed at twenty shillings, the Prior of
St Michael's Rock separately taxed at thirteen shillings and fourpence,
quite a high figure for an individual. The lists specifically mention the
monastery on the skeilic, not on the mainland, but the reference to a
prior rather than to an abbot suggests a subordinate role in an out-

station. Long after this, the head of the Ballinskelligs community was addressed in papal missives as the 'Augustinian Prior of St Michael's Roche'. It may well be that by the beginning of the fourteenth century, the skeilic had become a retreat centre for men whose principal work now lay before them on the mainland opposite.

115
Unanswered Questions

The rest of the story can be quickly told. Skellig Michael was an appendage of the Ballinskelligs community until 1578, when Munster rebelled against the English Crown in a dispute that was partly religious, partly to do with land titles, and very much to do with Queen Elizabeth's fear of being invaded by Catholic Spain. After putting down the revolt, she suppressed a number of religious houses that had been under the protection of the earls of Desmond, and had escaped the earlier dissolution of monasteries by Henry VIII. The skeilic passed into secular hands and remained with various owners until 1820, when it was bought by the Corporation for Preserving and Improving the Port of Dublin, which built a lighthouse high up on the western end of the skeilic to warn off Atlantic shipping. Unfortunately, no account had been taken of the thick fogs that can blanket the upper half of the skeilic, when there is total clarity at sea level. This meant that the light was often invisible to approaching ships when it was most needed, and so a second lighthouse was built much lower down, which continues to function today, its predecessor now reduced to a ruin 200 feet further up the cliff. It was the builders of

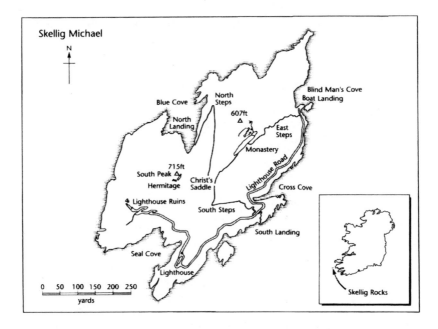

Skellig Michael

the lighthouse who were responsible for the metal projection above the summit of the South Peak, which has no religious significance, but was installed simply as a lightning conductor.

Although we do not know when the skeilic first became a place of pilgrimage, it was certainly celebrated as such while still in the possession of the Ballinskelligs community. Archbishop Dowdall of Armagh in the early sixteenth century called it one of the most important penitential stations in Ireland, and William Tirey, Bishop of Cork from 1623 to 1645, long afterwards recalled his own pilgrimage to the skeilic as one of the most memorable events of his life. But its heyday may have been over by the middle of the eighteenth century, when the *Ancient History of the Kingdom of Kerry* by a Muckross Franciscan was published, and referred to 'the great Skelike, formerly very much noted for pilgrimage over most parts of Europe'. In 1756 a certain Charles Smith, who described (inaccurately) an ascent of the South Peak without having visited Skellig Michael in person, remarked that 'Many persons about twenty years ago, came from the remotest parts of Ireland to perform these penances, but the zeal of such adventurous devotees hath been very much cooled of late.'

The penances may have included praying at Stations of the Cross en route to the monastery, and it is possible that one of the Stations— a severely eroded and roughly cruciform piece of stone—is still *in situ,* where the main staircase makes its first dog leg on the way up the south side of Skellig Michael. The most severe penitential act, however, was not only to scale the South Peak, but then to crawl out along the narrow ridge just below the summit which is known as the Spit, and at the end, with nothing but space around the penitent, to kiss an upright stone which had been deliberately lodged there in a socket, before inching back to the small summit platform and then beginning the descent. The stone is no longer there, having vanished (dislodged, presumably, in a storm) sometime in 1977.

One last question has arisen even more recently. We have never been sure how many monks dwelt on the skeilic at any period. At the time of the monastery's foundation, the custom was for new communities to be established following the pattern of Christ and his closest disciples; that is, with twelve men under a leader. But there has never been any suggestion of more than six clocháin on Skellig Michael, which would have meant a certain amount of doubling up if those figures held good, though more than one monk to a cell would have been contrary to all the precedents of the eremitical life. In the summer of 1994, however, in the course of some restoration work on Skellig Michael, archaeologists of the Office of Public Works (which is now the supervising authority, except for those portions of the skeilic which are controlled by the Commissioners of the Irish Lights) uncovered the foundations of a seventh beehive hut, whose existence had been quite unsuspected, in the middle of what was believed to be the monastic garden.

The question therefore is two-fold. Are yet more clocháin awaiting discovery? And where, then, did the monks grow their vegetables?

A BIBLIOGRAPHY

Abbreviations:

JRSAI   Journal of the Royal Society of Antiquaries of Ireland

DIAS   Dublin Institute for Advanced Studies

IHS   Irish Historical Studies

NMI   National Museum of Ireland

Alexander, J. J. G., *Insular Manuscripts, Sixth to Ninth Century* (Harvey Miller 1978)

Artz, Frederick B., *The Mind of the Middle Ages AD 200–1500* (Knopf 1953)

Attwater, Donald, *The Penguin Dictionary of Saints* (Penguin 1965)

Atwell, Robert, ed., *Spiritual Classics from the Early Church* (Church House 1995)

Backhouse, Janet, *The Lindisfarne Gospels* (Phaidon 1981)

Best, Richard I., ed., *Martyrology of Tallaght* (Henry Bradshaw Society 1931)

Bitel, Lisa M., *Isle of the Saints: Monastic Settlement and Christian Community in Early Ireland* (Ithaca, NY 1990)

Bonner, Gerald, Rollason, David and Stancliffe, Clare, eds., *St Cuthbert, His Cult and His Community to AD 1200* (Boydell 1989)

Bonwick, James, *Irish Druids and Old Irish Religions* (Dorset 1894/1986)

Bradley, Ian, *The Celtic Way* (Darton, Longman & Todd 1993)

Brown, Peter, *The Book of Kells* (Thames & Hudson 1985)

Cahill, Thomas, *How the Irish Saved Civilisation* (Doubleday 1995)

Carmichael, Alexander, *Carmina Gadelica: Hymns and Incantations* (Floris 1994)

Chadwick, Nora, *The Age of the Saints in the Early Celtic Church* (Oxford 1963)

———, *The Celts* (Penguin 1987)

Chadwick, Owen, *John Cassian: A Study in Primitive Monasticism* (Cambridge 1950)

Clark, Willene B. and McMunn, Meradith T., *Beasts and Birds of the Middle Ages* (University of Pennsylvania Press 1989)

Connolly, Hugh, *The Irish Penitentials* (Four Courts 1995)

Crone, G. R., *Early Maps of the British Isles AD 1000–1579* (Royal Geographical Society 1963)

Deanesly, Margaret, *A History of Early Medieval Europe from 476–911* (Methuen 1962)

de Paor, Liam, *A Survey of Sceilg Mhichíl* (JRSAI Vol. 85, 1955)

de Paor, Máire and Liam, *Early Christian Ireland* (Thames & Hudson 1958)

Delaney, Frank, *The Celts* (BBC/Hodder 1986)

Diringer, David, *The Illuminated Book: Its History and Production* (Faber 1967)

Edwards, Nancy, *The Archaeology of Early Medieval Ireland* (Batsford 1990)

Flaubert, Gustave (Kitty Mrosovsky, tr.), *The Temptation of St Antony* (Penguin 1983)

Gantz, Jeffrey, tr., *Early Irish Myths and Sagas* (Penguin 1981)

Gimpel, Jean, *The Medieval Machine: The Industrial Revolution of the Middle Ages* (Gollancz 1976)

Greene, David and O'Connor, Frank, eds., *A Golden Treasury of Irish Poetry AD 600–1200* (Macmillan 1967)

Greer, Rowan A., tr., *Origen: An Exhortation to Martyrdom . . .* (Paulist NY 1979)

Gregg, Robert C., tr., *Athanasius: The Life of Antony and the Letter to Marcellinus* (Paulist NY 1980)

Harbison, Peter, *Guide to the National Monuments in the Republic of Ireland* (Gill & Macmillan 1970)

———, *Pre-Christian Ireland* (Thames & Hudson 1988)

Hennessy, William M., tr., *Annals of Ulster: A Chronicle of Irish Affairs from AD 431 to AD 1540* (Dublin 1887)

Henschen, Folke (Joan Tate, tr.), *The History of Diseases* (Longman 1966)

Horn, Walter, Marshall, Jenny White and O'Rourke, Grellen D., *The Forgotten Hermitage of Skellig Michael* (University of California 1990)

Howlett, D. R., ed., *The Book of Letters of Saint Patrick the Bishop* (Four Courts 1994)

Hudson, Douglas Rennie, *Keltic Metal Work and Manuscript Illumination* (Metallurgia, April 1945)

Hyde, Douglas, *The Story of Early Gaelic Literature* (London 1895)

IHS, *The Arroasian Order in Medieval Ireland* (IHS Vol. IV No. 16 Sept. 1945)

Jackson, Kenneth H., *Studies in Early Celtic Nature Poetry* (Cambridge 1936)

———, tr., *A Celtic Miscellany* (Penguin 1971)

Jacobs, Joseph, *Celtic Fairy Tales* (Bodley Head 1970)

Jones, Gwyn, *A History of the Vikings* (Oxford 1968)

———, *The Legendary History of Olaf Tryggvason* (Jackson 1968)

King, A. A., *Liturgies of the Past* (Longman 1959)

Kinsella, Thomas, tr., *The Táin* (Oxford 1970)

Lavelle, Des, *The Skellig Story* (O'Brien 1993)

Little, George A., *Brendan the Navigator* (Dublin 1945)

Luibheid, Colm, tr., *John Cassian: Conferences* (Paulist NY 1985)

Lynch, P. J., *Some of the Antiquities Around St Finan's Bay, County Kerry* (JRSAI Vol. 32, 1902)

———, *Some of the Antiquities Around Ballinskelligs Bay, County Kerry* (JRSAI Vol. 32, 1902)

Mac Airt, Séan, ed., *The Annals of Inisfallen* (DIAS 1988)

Mac Cullagh, Richard, *The Irish Currach Folk* (Wolfhound 1992)

Maher, Michael, ed., *Irish Spirituality* (Veritas 1981)

McCann, Justin, OSB, ed., *The Rule of St Benedict* (Burns Oates 1963)

McCulloch, Florence, *Medieval Latin and French Bestiaries* (University of North Carolina 1962)

McLintock, H. F., *Old Irish and Highland Dress* (Dundalgan 1950)

McNeill, John T. and Gamer, Helena M., *Medieval Handbooks of Penance* (Octagon NY 1979)

Meyer, Kuno, ed., *Ancient Irish Poetry* (Constable 1913)

Meyer, Robert T., *Life of St Antony* (Longman 1950)

Mould, D. D. C. Pochin, *The Irish Saints* (Burns Oates 1964)

Murphy, Gerard, ed., *Early Irish Lyrics, Eighth to Twelfth Century* (Oxford 1956)

O'Floinn, Raghnall, *Irish Shrines & Reliquaries of the Middle Ages* (NMI 1994)

O'Meara, John J., tr., *Gerald of Wales: The History and Topography of Ireland* (Penguin 1982)

O'Donoghue, Noel Dermot, *The Mountain Behind the Mountain* (Clark 1993)

O'Donovan, John, ed., *Annals of the Kingdom of Ireland by the Four Masters from the earliest period to the year 1616* (Dublin 1851)

Payne, Ann, *Medieval Beasts* (British Library 1990)

Phillips, Walter Alison, *History of the Church of Ireland;* Vol. 1, *The Celtic Church* (Oxford 1933)

Reeves, William, *The Culdees of the British Islands* (Dublin 1864)

Renfrew, Colin, *Archaeology and Language: The Puzzle of Indo-European Origins* (Cape 1987)

Ríordáin, Seán P., *Antiquities of the Irish Countryside* (Methuen 1979)

Ryan, John, SJ, *Irish Monasticism; Origins and Early Development* (Four Courts 1931/1992)

Ryan, Michael, *Early Irish Communion Vessels* (NMI 1985)

Savage, Anne, tr., *The Anglo-Saxon Chronicles* (Heinemann 1983)

Selmer, Carl, ed., *Navigatio Sancti Brendani Abbatis from Early Latin Manuscripts* (University of Notre Dame, Indiana 1959)

Seznec, Jean, *The Temptation of St Antony in Art* (*Magazine of Art* Vol. 40 No. 3 Washington DC 1947)

Sharpe, Richard, ed., *Adomnán of Iona: Life of St Columba* (Penguin 1995)

Sherley-Price, E., tr., *Bede: A History of the English Church and People* (Penguin 1965)

Stokes, George T., *Ireland and the Celtic Church* (Hodder & Stoughton 1887)

Stokes, Whitley, ed., *The Martyrology of Oengus the Culdee* (London 1905/Henry Bradshaw Society 1984)

Ward, Benedicta, SLG and Russell, Norman, *The Lives of the Desert Fathers* (Mowbray 1983)

Went, A. J., *Irish Fishing Spears* (JRSAI Vol. 82, 1952)

Wilkinson, George, *Practical Geology and Ancient Architecture of Ireland* (Murray 1845)